ABOUT THE AUTHOR

Photo by Monty Coles

Tanya Ha is an award-winning author, science journalist, television presenter and environmental campaigner. A well-known figure in the Australian green movement, she is often described as the 'people's environmentalist'. Tanya is a firm believer that the environment is relevant and important to everyone, and that ordinary people can achieve great things when they're made to feel empowered, rather than guilty.

Tanya is currently a science reporter for ABC1's flagship science show *Catalyst* and is also the resident eco expert on *Can We Help?* She also featured in the award-winning SBS television show *Eco House Challenge* as the show's resident eco-coach and was the host and main reporter of *Warm TV* (WIN Television), for which she won the United Nations Association of Australia World Environment Day Awards 2010 Media Award for Environmental Reporting.

Tanya spent seven years working for the green group Planet Ark and continues to support and assist the work of a range of other environmental organisations. She also serves on the boards of the state government authority Sustainability Victoria, is a past board member of the green group Keep Australia Beautiful and an ambassador for the innovative WA Government's Living Smart program. In 2008, Tanya was one of the 1000 notable Australians selected to participate in the Australia 2020 Summit (Sustainability and Climate Change stream).

Tanya's first book, the best-selling *Greeniology*, has been published in Canada and translated into French, with a Chinese language edition in the works. Her second book, *The Australian Green Consumer Guide*, was published in late 2007 to rave reviews, followed by the 'Greeniology' series of pocket-sized guides. 2009 saw her turn her attention to younger readers, with publication of the acclaimed *Green Stuff for Kids*. She writes feature articles and columns for several magazines and newspapers, including the long-running 'Ask Tanya' green living advice column for *G magazine*, and blogs on science and sustainability on the *Catalyst* show's website.

Tanya lives in Melbourne with her husband and two children.

PRAISE FOR TANYA HA'S PREVIOUS BOOKS

Greeniology

'A must for all those who want to do their bit for the planet. Al Gore would approve.'
—*The Age*

'Tanya Ha has done a wonderful job with this book and should be congratulated. *Greeniology* provides practical advice on what we can do to lessen the load on the planet. An informative and entertaining read.'—*Gardening Australia*

'A delightful, accessible treasure trove of a book ... and never more relevant.'
—Robyn Williams, *The Science Show*

'Green is not just for the garden. Every household needs a copy of Greeniology.'
—Jamie Durie

'I want to see our beautiful planet preserved for our children and our children's children. Tanya Ha shows people how to live their lives in a more sustainable way—every woman should read this book.'—Olivia Newton-John

'Her 'light green' approach is all about positive things everyone can do to help the environment. Subtitled 'how to live well, be green and make a difference', Greeniology is packed with well-researched, practical information on topics ranging from disposable nappies to solar power.'—*Sydney Morning Herald*

'Environmental spokesbabe Tanya Ha's *Greeniology* is a cute but comprehensive ecofriendliness primer.'—*Montreal Mirror*

'The great value of *Greeniology* is in the plethora of very easy alternatives that it presents, combined with an empowering emphasis on our ability to make choices.'
—*Journal of Australian Studies*

'This book should be a slightly numbing account of how you can be more environmentally conscious, but instead it is very appealing. Simple things count, like saving water, recycling and smart buying. It all adds up to saving money as well as helping the planet. Short, snappy tips and explanations help keep the pages turning.'—*Daily Telegraph*

The Australian Green Consumer Guide

'Tanya Ha's straightforward approach to green shopping is simple to read and gently persuasive.'—*Green Magazine*

Green Stuff for Kids

'Something fun, informative and challenging for green kids (and their parents) to sink their teeth into.'—Tim Flannery

'You can make a difference and this clearly set-out book is an excellent guide.'
—*West Australian*

'Rather than simply presenting cold facts, Ha's commentary is interesting and at times humourous' ... 'superb advice' ... 'highly recommended' ...—*Sun Herald*

'While Ha does a great job of explaining the facts, the designers should be congratulated for creating such a user-friendly book (graphs, break-outs, check-lists, projects) that makes absorbing Ha's engaging writing style even easier ... Invaluable reading for those aged 10 and up.'—*The Sunday Age*

'... a goldmine for teachers and kids alike.'—*G magazine*

Greeniology
2020
Greener living today, and in the future

Tanya Ha

MELBOURNE
UNIVERSITY
PRESS

Dedicated to Archer, who by 2020 will be on the doorstep of adulthood

MELBOURNE UNIVERSITY PRESS
An imprint of Melbourne University Publishing Limited
187 Grattan Street, Carlton, Victoria 3053, Australia
mup-info@unimelb.edu.au
www.mup.com.au

First published 2011
Text © Tanya Ha, 2011
Design and typography © Melbourne University Publishing Ltd, 2011

'Ask Tanya' questions and answers appear courtesy of nextmedia Pty Ltd, publishers of
G magazine.

Illustrations by Andrew Treloar and Tanya Ha
Typeset by TypeSkill
Printed by Griffin Press, South Australia

National Library of Australia Cataloguing-in-Publication entry

Ha, Tanya, 1972–

Greeniology 2020: greener living today, and in the future/Tanya Ha.

9780522858549 (pbk.)
9780522860511 (ebook)
Includes index.

Household ecology.
Environmental protection—Citizen participation.

363.70525

Cover image: The plant pictured is type of pennywort, which is often
grown for medicinal use. It's included here as a symbol of the relationship
between human health and the environment. However, it can be weedy,
so if you grow it at home, make sure it stays confined.

Contents

Introduction

Have you ever seen a news story about the thousands of hectares of rainforest being felled each year to provide grazing land for hamburger cattle? Have you heard that we're seeing the extinction of species at the fastest rate ever, and that the air is filling with greenhouse gases and turning the earth into a toaster oven? Birds are dying because they're covered with oil; ships carrying nuclear waste are probably trying to dock at your local port, and some Pacific islands may well be flooded by rising ocean levels. It's not pretty, and it's all our fault!

You probably care about these issues, but can't imagine how you can stop any of it happening. You can't see yourself going on a hunger strike or being chained to a bulldozer in protest. You're not alone. While there are a few hard-core activists at the frontline of the environmental movement who are not afraid to further the green cause through extreme and very public acts, this approach is not for everyone.

Public protest was (and still is) necessary. Someone had to bring the plight of the environment to the attention of the average person and to make us understand that human activities haven't been the best thing for all creatures great and small.

But I've got good news: protesting isn't the only way to help save the planet. There's a new wave in the environmental movement, defining itself by what it's for rather than what it's against. The new environmentalist knows that it's not just the fault of governments and big businesses, and that we all share the blame for damaging the planet. After all, we purchased the products that harmed the earth through their manufacture or disposal. We bought and used the hairsprays that contained ozone-destroying CFCs. Perhaps we even wore fur and ate shark's fin soup. Our generation has made many environmental mistakes innocently. The way to move forward is to take back responsibility for looking after the planet—to stop complaining and to start concentrating on being part of the solution, rather than just feeling guilty.

So who is part of this new wave of environmentalism? It's hard to tell because they look just like ordinary people. They *are* ordinary people, with jobs, friends, families and busy lives—people who happen to be very far-sighted and who want to preserve this beautiful planet and enjoy it for years to come. They want to enjoy the best and most comfortable lifestyle possible but are also determined to balance their needs with those of the planet.

Rather than waste time just sitting around and talking about it, they want to get up and do something, by changing their everyday lives, by not using the products that cause harm and by supporting those that are greener. They draw the line at excessive consumption. Their clothes may not be made from hemp, but they'll wash them with a greener detergent in an energy- and water-efficient washing machine, powered by green electricity from GreenPower-approved sources. They're not heroes, ready to scale tall buildings. They're not selfless saints, making sacrifices for their cause. They believe in 'working smarter' rather than working harder. They want there to be enough resources for their children and grandchildren. They may have heard 2010–20 referred to as the 'Transition Decade', so they sense the urgent need for more action and less 'talk'. They have a vision for the future, and they want to make headway today.

Greeniology 2020 is about living green today with a vision for the future. It shows you how to change your life to make it greener, and helps you to understand why those changes matter. It's the science and art of living a greener life. There are so many things that we can all do to preserve the planet: saving energy and water in our homes, recycling, buying wisely and using less paper at work. These actions may seem like small things, but if we target our efforts where they will be effective and keep our eyes on our goals, we will achieve a huge amount.

This book is all about action, but it's also helpful to be better informed about the major environmental issues we're facing—the 'hot topics' that we hear about in the media and that drive the need for eco-action. If you're interested in this kind of information as well as action, I've provided a downloadable overview of the environmental hot topics that inform the advice in this book, online at www.tanyaha.com. Many of these topics are also covered in a fun and informative way in the book *Green Stuff for Kids*. Don't be put off by the title if you're an adult. Although I wrote *Green Stuff for Kids* for a younger audience, there's plenty in it for grown-ups too.

HOW TO READ THIS BOOK

Chapter 1 (on 'Becoming future ready') sets the scene for becoming greener during the Transition Decade. It not only contains some cold, hard truths about what to expect in coming years, but also describes the kind of strategic approach that will help you to achieve results quickly. If you would rather cut straight to the action, then move on to later chapters, which cover the various aspects of green living, from saving energy and water to building houses and buying cosmetics.

The chapters themselves are themed around goals or action areas, and the advice within them is informed by the many years I have worked in the field, running sustainability workshops for local councils, small to medium businesses, mothers' groups, community groups and others. If you want to reduce your energy bills, go to chapter 9 ('Energy'); if you're concerned about the health of your family, start with chapter 2 ('Healthy home, healthy planet'); take chapter 3 ('Working green') to the office, and so on.

Start today! Then each week, try to change at least one habit in your life to a greener alternative. Set goals and use the worksheets in the book as a reminder and checklist to monitor your progress. It's easy, and the world will reap the benefits of your efforts.

 This symbol refers you to another page or to a website for more information on a particular topic.

 Where you see this symbol, follow the green advice to save money.

 This symbol shows you how to improve your health while caring for the environment.

1

Becoming future ready

Change has a considerable psychological impact on the human mind.
To the fearful it is threatening because it means that things may get worse.
To the hopeful it is encouraging because things may get better. To the
confident it is inspiring because the challenge exists to make things better.

King Whitney Jr

If you believe all the gloom and doom in the media, the future looks dark and grim, like a scene out of the movie *Blade Runner*. I have a different mental picture of the future. To me, the future is green—literally. We will have blurred the line between city and country, bringing life back into our cities, with trees to provide shade, vegetated roofs to provide insulation, fruit trees on nature strips, and vegie patches in backyards. I've seen the beginnings of it. A hundred years ago, Swanston Street in the centre of Melbourne was a grey, building-lined street, dominated by the sombre tones of local bluestone. It's now lined with lush, green plane trees.

This greening of cities is catching on, and not because of some caring, sharing, tree-hugging sentiment. It's growing because urban planners are realising that plants combat the urban heat island effect. They also know that urban gardens and rooftop vegetation can help control stormwater and flash flooding. And they've seen studies that show that gardens and plants in hospitals can lead to better medical outcomes and reduced health costs.

Humans are resilient, innovative and adaptable creatures, which is why I have such hope for the future. Change is coming for a number of environmental, social and political reasons. It's important that we snap out of a 'victim' mentality and face the future with open eyes and a determination to meet the challenges ahead.

This chapter provides an overview of current and future environmental challenges. Before plunging into the advice and tips for action that form the vast bulk of this book, we need to understand that aspects of our environment and society will inevitably change, such as weather patterns or the price of electricity, and these changes will affect our living habits and our ability to shape them. We will need to plan the changes we're going to make to our lives so that we get the greatest result for our effort and investment. In sporting terms, this chapter is all about examining the playing conditions and coming up with a game plan.

WHAT DOES THE FUTURE HOLD?

While we may not know the exact number of centimetres that sea levels will rise, if mobile phones will be made without tantalum or when peak oil will be reached, there are a few things that we do know are likely to happen. At first glance, they may seem depressing, but keep in mind that humankind has survived world wars and other traumatic challenges over its history, and has gone on to thrive.

Increasing costs

The main factors that are driving up prices are Australia's growing population, competing demands for resources worldwide, conflict in oil-producing countries, and changing environmental conditions, including climate change.

Electricity prices have already begun rising in much of Australia. In 2010 the Minister for Resources and Energy, Martin Ferguson, told a meeting of energy supply CEOs that Australia needs $100 billion worth of investment in electricity infrastructure over the next decade just to meet growing demand and to replace ageing infrastructure. A price on carbon, if introduced, will have a relatively small influence on price, particularly compared with the high capital costs of electricity infrastructure. Households and businesses with solar panels will be cushioned from these price increases. The households least affected will be those that are lean and green and don't waste electricity. While increasing living costs are naturally of concern, keep in mind that Australia has had the luxury (by world standards) of cheap electricity for many years—residents in Germany pay more than double the average Australian tariff, and Danish households pay triple.

Water prices will rise for similar reasons. As well as the increasing costs of supplying the demands of a growing population and replacing ageing infrastructure, climate change research predicts less rainfall overall. So Australia's more water-scarce population centres are safeguarding against future drought by investing in desalination—a highly expensive backup plan—and in other alternative freshwater sources.

Petrol prices have risen, fallen and risen again. Some experts put the temporary fall down to the economic downturn that followed the global financial crisis. But oil prices will continue to rise due to increasing demand from growing economies in developing nations such as China, India and Brazil. Alongside this growing demand are troubles in the regions that supply us with oil. Wars in the Middle East in recent decades—along with more recent conflict and political instability stemming from the wave of protests in the Middle East and North Africa in 2010–11—have seen sharp rises in the price of oil. This is reflected in the price at the pump when we refill our cars.

The size of the hit to your hip pocket is a product of two things—the price per litre and the number of litres you use—so the impact of these cost increases depends, to some degree, on how much 'gas' we guzzle. It's a 'user pays' system. The good news is that we can insulate our hip pockets against the affects of these price rises by reigning in our consumption. This book is packed with ways you can reduce your consumption of electricity (and energy in general), water and petrol.

A price on carbon

Whether through a carbon tax or a permit-trading system, putting a price on carbon is intended to make polluting technologies and activities more expensive. This creates a marketplace that is more favourable to the development of less-polluting alternatives. Some experts argue that putting a price on carbon is really a way of making the price of fossil fuels reflect their true costs, and of making the user pay. For example, when we pay for petrol, we don't directly pay for the healthcare costs of people who have respiratory illnesses exacerbated by smog, but all taxpayers ultimately pay for our public-health system. There will be health consequences from climate change, so a cost to the major contributors to climate change seems a fitting way to meet these future health costs. If they don't, who will foot the bill?

A price on carbon will increase the price of electricity derived from fossil fuels and the cost of products that have a large carbon footprint. Whether that will increase your living costs depends on you. Ideally, it will make people think twice before wasting energy, encouraging more energy-efficient behaviour. If that's the case, then increases in the unit price of electricity can be offset by using less, leaving us no worse off. Products that are made with energy-intensive materials will theoretically become more expensive, but manufacturers who want to maintain their market share will have an incentive to make their products using different materials so that they can offer them for the same price. In short, a carbon price will change the marketplace to favour greener alternatives, but it won't be the end of the world.

A changing climate

Climate change has some serious consequences, some of which we're already experiencing. 'Climate change adaptation' is the term used to describe measures we take to reduce our vulnerability to expected climate change impacts. Some things we can safeguard against at a household level; others require action from government. Australia is a large country with many different climate zones, so climate change will have different consequences depending on the region. (The Climate Change in Australia website provides a detailed look at how climate change will affect Australia across its many regions— see http://climatechangeinaustralia.com.au.) The following are some of the extreme events and increased risks we're expecting to see or are already experiencing as a result of climate change.

Temperature increases All of Australia is expected to see rising average temperatures, with stronger warming inland. This includes an increase in the frequency of days over 35°C and of heatwaves, particularly in Victoria and South Australia. Heatwaves are particularly severe in urban areas, where the thermal mass of buildings and pavement absorbs and stores the heat in what's known as the urban heat island effect. Heatwaves can lead to deaths and illness from heat stroke. In fact, the extreme heat in Melbourne in the lead-up to the Black Saturday bushfires is estimated to have caused 374 deaths above the normal mortality rate, eclipsing the 173 deaths from the fires themselves. It is vital, then, to provide proper shading to prevent heat gain (see pages 140–1). You can also landscape to provide shade and the natural airconditioning effect of vegetation (see pages 218–21).

Changing rainfall patterns Decreases in rainfall are likely in the southern parts of Australia, while increases in rainfall are possible for the Top End. Some areas are expected to experience less frequent but more intense rainfall episodes, which can damage and erode soil and cause flash flooding. This is particularly problematic in urban areas where housing density has increased without capacity improvements to stormwater drainage systems. For example, the City of Port Phillip in Melbourne lies on the coast and has seen a shift over the last few decades to higher-density housing because of high land values and easy access to the city. Suburban blocks that once had a house with a garden that soaked up rainwater are now being replaced with multi-unit dwellings that almost completely cover the ground with roof or concrete. All of the rainfall on these developments flows into stormwater drains, and the region has less garden space to soak up some of the stormwater. Consequently, some areas are prone to flooding. This may be an emerging problem in your suburb. The benefits of rainwater tanks in areas like these go beyond just providing an alternative source of water.

Drought Decreased rainfall and declines in soil moisture produce drought, which affects the livelihood of farmers, the price of food and the health of ecosystems. It can also kill trees in urban settings, which may then be more vulnerable to high winds and storm damage. Keep this in mind if you have large trees on your property. Have them checked by a tree surgeon, particularly after periods of prolonged drought. The areas most at risk are southern and south-western Australia.

Bushfire Heatwaves, low rainfall, high winds and drought combine to create a perfect storm in terms of bushfire risk. Find out if you live in a high-risk area and develop a bushfire plan. Take particular care with landscaping.

Rising sea levels Sea levels are expected to rise gradually, due to increased water from melting icecaps and to 'thermal expansion' (the way that materials expand slightly when warmed). Coastal areas are at risk, particularly from storm surges caused by intense weather systems.

Shifts in climate patterns Changing climate patterns are seeing tropical conditions shift south. This is changing the distribution of certain species, such as fish stocks, agricultural pests and disease-carrying mosquitoes. We're also seeing a shift in a storm belt that occasionally brings hail closer to Sydney. The Sydney hailstorm of 1999 caused more than $2 billion in damage. Changes in storm patterns are worth thinking about in relation to your roof materials and structure if you live in storm-prone areas.

Changes in alpine regions As Australia warms, we will also see decreases in snow cover, changes in the length and timing of ski seasons, and pressure on alpine plant and animal species.

More intense storm activity Warm oceans result in more evaporation, and in storms and cyclones that carry more water. While it is difficult (and somewhat pointless) to determine whether climate change caused Cyclone Yasi, it is fairly certain that climate change increased its intensity and the damage it caused. The lesson from Cyclone Yasi is to avoid building on flood plains and close to rivers. The Bureau of Meteorology and Emergency Management Australia have together produced an information package on severe thunderstorms. For more information, see www.bom.gov.au/info/thunder.

The Bureau of Meteorology website also contains current storm warnings, so it is a useful website to have bookmarked if you live in an area at risk.

HOW DO I BECOME FUTURE READY?

Now that we know the worst of it, what can we do about it? Plenty! But first we need the right mindset. New technology can't save us if we refuse to try it. Behaviour change has its roots in psychology and in an understanding of how we think, what motivates us and what makes us truly happy.

Understand your values, goals and motivations

Having a conscience is part of being human. Most of us avoid deliberately hurting others, and have a sense of right and wrong and what is fair. But we are living in an unfair and unequal world. The environment isn't helped by our feelings of guilt over being lucky enough to live in a land like Australia. Too much guilt is debilitating. But we can have values that motivate us to change our lifestyles so that we can live in better balance with the natural environment. Your values may include:

- a respect for the rights and welfare of animals
- a concern for your personal health or the health and wellbeing of your children
- a resourceful mindset that hates waste
- a strong sense of responsibility
- a deep love of the natural environment and the beauty of nature
- a high regard for simple living, or a desire to escape the 'rat race'
- a desire for independence and self-sufficiency
- an intellectual interest in sustainability issues and a desire for smarter living habits.

Understand and acknowledge what's important to you, so that you can make choices that align with your values. Remember that your values and priorities may be quite different from those of your neighbour. Many sections of this book offer a suite of tips and options, rather than an inflexible list of instructions to be universally obeyed. Choose the options that resonate with you. Always keep in mind the vision of what you want to achieve through greener living. When someone asks you why you're making an effort, this will be your answer.

Get strategic

Enough awareness-raising! It's time to make some serious inroads into reducing our footprint on the earth. First identify the parts of your life that are high-consuming and act there. One way of looking at it is to compare it with dieting. You don't lose weight by ordering a skinny latte to go with your large slice of cheesecake. Target the environmental 'cheesecakes' of your life to get the quickest results. Yes, the little things will add up, but the big things add up faster.

A good way to go about this is to work out action plans for the areas of your life you want to work on. These may include an 'Energy Action Plan', with the ultimate goal of becoming carbon neutral with your use of household energy, or a 'Waste Action Plan' to make your garbage bin a little less full each week. A typical action plan has the following steps:

1 **Outline the goal**. It doesn't have to be specific. It could be as simple as cleaning up your cleaning routine or reducing the amount of waste you send to landfill, or as complex as a home renovation.
2 **Measure your current situation**. This involves identifying your current level of environmental impact. If your goal is to reduce your waste, then your measure may be a simple 'guesstimate' of how full your bin is each week. For electricity, gas and water, your bills measure your current level of consumption.
3 **Think about where there may be easy opportunities for change**. This step is a little like a SWOT analysis. Look at your home circumstances and identify the strengths, weaknesses, opportunities and threats. For example, your old-fashioned showerhead may waste a lot of water, but on the bright side, changing it to a water-efficient model is a 15-minute DIY job that will make a big difference to your water use.
4 **Make the changes**. Apply the tips and ideas in this book to reduce the impact that you have identified. Some of these things will be easy to achieve, such as adjusting the thermostat on the heating—a simple habit change. Others might take a moderate level of effort, such as changing light bulbs or applying mulch in the garden. Others still are more occasional opportunities, such as replacing whitegoods. Plan the changes you will make in the short term, medium term and long term.
5 **Look for alternatives**. In the case of energy and water, once you've reduced your consumption you should also consider other ways to provide those needs with less impact, such as using solar panels as an alternative source of energy or a rainwater tank as an alternative water source.
6 **Compensate or offset**. In some cases, there are things you can do to offset your environmental impact.

Make a commitment

This is where your willpower and perseverance come into play. We've all made New Year's resolutions, but how many of us keep them? Psychologists say we're more likely to make changes if we write them down or make a public commitment to them. Social media– and internet-based campaigns such as 1 Million Women (www.1millionwomen.com.au) offer a virtual 'public place' to make these commitments and to share ideas.

For some people, it's also helpful to write a 'To do' list of actions. There's something particularly satisfying about checking them off with a big, fat tick as each item is met. At the end of each chapter, starting with chapter 2, there is space for you to write down your commitments to yourself. An example appears below. Write them down, tick them off as you accomplish them, revisit your commitments every now and then and feel free to make some new ones.

MAKE A COMMITMENT

This week I will...

☐ _____

☐ _____

☐ _____

This month I will...

☐ _____

☐ _____

☐ _____

When I get the opportunity I will...

☐ _____

☐ _____

☐ _____

Record your results and keep up the effort

Keep track of your efforts to reduce your environmental impact by monitoring your electricity, gas and water bill readings and, if you have a car, take regular odometer readings. Keep a record of them in your diary or in a spreadsheet on your computer so that they're easy to find when you want to make a comparison. Or download and use the tables included with other *Greeniology 2020* resources online at www.tanyaha.com. Aim for continuous improvement.

Habits can take a while to change, but with persistence you will get there. Before you know it, greener living will feel normal and effortless. The rewards will be a healthier place to live, a home that's cheaper to run and more comfortable to live in, and that quiet inner peace that tells you you've faced up to our society's environmental problems and you've made an effort to be part of the solution. As Mahatma Gandhi observed, 'You must be the change you wish to see in the world.'

2

Healthy home, healthy planet

For some people, the motivation for going green is to benefit the broader environment: to cut pollution, fight climate change and reduce the harm we perhaps inadvertently inflict on the earth's many species. Others are motivated by their love and concern for one species in particular—humanity—and for particular members of that species, such as their children, friends and families. Both of these motivations are beautiful and admirable.

If your interest in helping the environment stems from concerns about human health, then this particular chapter is for you. The health-related aspects of green living are touched on throughout this book, but this chapter collects a few other issues that overlap with several chapters and are worth special attention. Aside from my intellectual interest in health issues, I write this from the perspective of having been a childhood eczema sufferer, from having watched my son at the age of three sitting in a big hospital bed struggling with asthma, and from having lost loved ones to cancer. To those who resist the call to live in better balance with the natural environment for reasons of self-interest, I would point out that it's not all about polar bears, saving whales and hugging trees. Sustainability is really about safeguarding the future of our species and recognising that we're part of nature, not separate from it. One of the many benefits of greener living, aside from saving money and living in homes that are not too hot or cold, is that it's better for our health. This chapter focuses on indoor air quality and endocrine disruptors. Other chapters that are particularly relevant to health and well-being are those on 'Working green' (chapter 3), 'Green cleaning' (chapter 4), 'Green grooming and bathing' (chapter 6), 'Food' (chapter 7) and 'Green building and renovating' (chapter 12).

From the outset, let's note that 'chemophobia'—the irrational fear of chemicals—isn't helpful. Some of our phobias are really fear of the unknown. The antidote is knowledge, so this chapter aims to provide information so that you can make informed choices, based on your own household's needs and sensitivities. Finally, keep in mind that the information here is general.

How relevant it is to your situation will depend on your own body and its strengths, weaknesses and sensitivities. Keep the information in mind, but remember that the best place to get health advice tailored to your needs is from a trained medical professional who has seen you and examined you. If you have a health issue, seeing a doctor is far more useful than reading a book.

INDOOR AIR QUALITY

Recent years have seen marked increases in asthma, allergies and multiple chemical sensitivity in people of all ages. Our modern lifestyle exposes us to a huge range of substances, while the hours we spend indoors result in constant low-level exposure to indoor air pollutants. It's important that we protect our health by taking steps to provide clean air to breathe inside our homes, particularly since we spend about 90% of our time indoors. You can do this in three steps:

1 Prevent indoor air pollution.
2 Ventilate to remove pollutants.
3 Remove excess moisture.

Types of indoor air pollution

Indoor air pollution comes from a variety of sources. You can minimise the contribution of each of them.

Health impacts of air quality

According to CSIRO, the number of deaths due to air pollution in Australia is higher than the road toll. An estimated 2400 deaths each year are linked to air quality issues, which equates to a death every four hours. This death toll increases if the long-term effects of air pollutants on cancer are included.

According to the Department of Sustainability, Environment, Water, Population and Communities, the health issues linked with exposure to poor indoor air quality include: skin, eye and throat irritation; headaches; drowsiness; hypersensitivity; and odour and taste symptoms. Long-term exposure to chemicals such as formaldehyde may cause cancer or respiratory disease.

Combustion gases from gas, propane, oil, kerosene and LPG heating and cooking can be serious indoor air pollutants. Carbon monoxide poisoning from car exhaust is well known, but in rare circumstances it can be a killer in the home. In Victoria in 2010, two boys died and their mother has continuing illness from carbon monoxide poisoning from a faulty gas heater flue. In early 2011, a man died from asphyxiation after taking shelter from Cyclone Yasi in an enclosed room with a diesel generator running to provide power. Don't use unflued gas or kerosene heaters, and have the flues of other gas heaters checked periodically to ensure they're functioning correctly. Generators should never be used in enclosed spaces.

Cooking is one of the surprising sources of indoor air pollutants. The simple way to prevent some of this is to avoid burning food (easier for some cooks than others!). However, even without burning, there are still some combustion pollutants and unburnt gas released, so always use a fan when cooking.

Cigarette smoke that is unfiltered and passively smoked is more polluted than the smoke drawn through the filter by the smoker. Smoking is bad enough for your health on its own. Avoid smoking in enclosed spaces or you'll passively smoke the remaining fumes. Smoking in bed can also be a cause of household fires.

Gas pollutants are emitted by dry-cleaned clothes and by the materials that some beds, furniture and mattresses are made from. Carpet, plastic products, clothes, books and other printed materials, foam insulation and small electrical goods can also release gas pollutants, including volatile organic compounds (VOCs), such as formaldehyde. In particular, try to minimise the amount of these products you have in the room where you sleep. New homes, renovated or freshly painted rooms, or those containing new furniture or other products that emit gas should be thoroughly ventilated as often as possible for the first few months. Some building materials release a lot more VOCs and other pollutants than other material types, so factor this into your choices when building or renovating.

Toiletries, perfumes and cleaners may sometimes smell nice, but they're not always good for you. This includes perfumed toiletries and spray deodorants. Choose personal care products and cleaners with care, and follow the safety instructions of cleaning products.

Paints, pesticides, solvents, art and craft materials and DIY products can all emit potentially harmful fumes. Some hobbies should be done outside the house, perhaps in a garage, shed or workshop. If you do a lot of craft or DIY, consider having an exhaust fan installed in your shed, workshop or garage. Also use ventilator masks while working. They are available from hardware stores and can include a range of different replaceable filters depending on the type of pollutant you wish to remove from the air. With pesticides, prevention is better than cure, so eating at the dinner table will prevent pest-attracting foods from being spilt in bedrooms. Avoid using pesticides in bedrooms at all costs. All of these products, particularly those that contain volatile ingredients, have the potential to emit pollutants even when they're not in use, particularly in hot weather. Store them in a cool, dry and dark place in the garage or a storage shed outside of the house.

Airborne particles, particularly fine solid ones, can get into your lungs and irritate them. Take care when using powdered products or dusting.

Natural pollutants are biological sources of irritants, such as dust mites, moulds and other fungi. Mould and fungi can be avoided by preventing a build-up of moisture inside. If you find that condensation forms on your windows on cold mornings, there's a good chance that your room isn't adequately ventilated. Make sure your home has adequate ventilation, and clean away any patches of mould that start to form.

Pets, along with their hair and their bedding, can contribute to poor air quality, with the consequences ranging from simple bad smells to severe allergic reactions. For most people it is healthiest to keep pets away from the bedroom. Keep your pets clean, and vacuum the house regularly. Animal lovers with allergies may unfortunately have to love their animal friends from afar and give up pet ownership.

Dust mites are tiny insects that thrive in warm, moist conditions and live on a diet of dead skin cells. In the house, they're most commonly found in beds, bedding and carpets. They can also live in curtains, couches, cushions and even soft toys.

The dust mites themselves aren't much of a problem. Their fine faecal matter, however, readily becomes airborne, and when inhaled by humans it can trigger a strong allergic reaction. Dust mites particularly cause problems for some asthmatics. For information on how to control dust mites, see chapter 4, page 57.

Factors that influence air quality

- **Living habits** Some indoor activities elevate the levels of some pollutants. These activities include cooking, using candles,

Healthy eco-labels

Good Environmental Choice Australia or GECA (www.geca.org.au) is an environmental labelling program that certifies products that have met certain environmental and health standards based on an assessment of the entire life cycle of the product, from manufacture to disposal or recycling. Product criteria include restrictions on ingredients that contribute to poor indoor air quality. The program covers a wide range of products, including paints, timber treatments and varnishes, cleaning products, furniture, outdoor furniture, Masonite hardboard, plasterboard, flooring, insulation, recycled plastic products, textiles, carpet, waterless car wash products and adhesives. Look for products bearing the GECA logo.

Sensitive Choice (www.sensitivechoice.com.au) is a labelling program run by the National Asthma Council Australia. The logo is awarded to products that have been reviewed by an advisory panel and found to offer a better choice to asthma and allergy sufferers. While not strictly an environmental labelling program, products that are suitable for allergy sufferers tend to have less environmental impact, so the program may be useful as a guide.

using open fires, smoking, using aerosol deodorants and cleaning products, and using nail varnish.

- **Season** There are seasonal variations in the levels of some pollutants. For example, we're more likely to open windows to ventilate in warmer weather, while in cooler weather we're more likely to be exposed to combustion fumes from heating or cooking hot meals.
- **Moisture** A moisture problem can promote the growth of mould and other fungi.
- **Hobbies** Some art and craft materials release air-polluting fumes.
- **The presence of off-gassing products and materials** 'Off-gassing' or 'out-gassing' is a process in which a solid releases gas. For example, formaldehyde can off-gas from composite board or plywood made with formaldehyde-containing adhesives. Off-gassing is responsible for 'that new smell' that some products have.
- **The age of the home** Older homes tend to be leakier; newer homes tend to have more plastics and composite materials that off-gas, and they are more airtight.
- **Internal garage access** A door connecting a garage to the house provides a pathway for pollutants to enter the house.
- **Exterior air quality** Busy roads or local industry may mean that the outdoor air isn't always 'fresh'.

Plants that clean the air

Bring the outdoors indoors! It's good feng shui and will help to make your sleep sweeter and healthier. All of these plants help to clean the air, removing pollutants such as formaldehyde, n-hexane, benzene and trichloroethylene:

- aloe vera
- happy plant (*Dracaena* varieties)
- peace lily (*Spathiphyllum* varieties)
- gerbera
- chrysanthemum
- kentia palm (*Howea forsteriana*)
- butterfly or golden cane palm (*Chrysalidocarpus lutescens*)
- spider plant (*Chlorophytum comosum*)
- rubber plant
- aspidistra
- Boston fern.

Ventilating to remove pollutants

Ventilation is the elimination of stale, moist air and indoor air pollutants from inside the house, replacing it with 'fresh' air from outside the house. In warmer climates, ventilation can also provide free cooling. In Australia's cooler southern states, there is a push to make our homes more airtight in an effort to reduce heat loss in winter. However, we need to make sure that we don't create an indoor air quality problem in our quest for improved energy efficiency. Cutting drafts from outdoors reduces a passive source of fresh air, making ventilation more important. Homes with good ventilation can still be efficient and can also help to minimise moisture problems. Here is an overview of the most common types of ventilation.

Passive ventilation (or natural ventilation) is as simple as leaving one or more windows open for a short time and allowing the movement of air into and out of the house. This is not always effective or economical, particularly in winter, as it lets heat out of the house and cold air in, increasing the heating load. Any savings made by avoiding the purchase and running of a mechanical ventilation system are soon outweighed by the increased heating costs.

Natural ventilation is best incorporated into a house's design, particularly in tropical and hot, dry climates. A house can be sited to catch cooling breezes, and windows can be placed to allow easy cross-ventilation. In particular, houses near the coast can be designed to catch the afternoon sea breezes in summer.

Exhaust-only ventilation uses exhaust fans to draw air from a specific location and vent it outside. They are good for removing air from a localised and concentrated source of moisture, smells or pollutants such as the shower, toilet or stove.

Heat recovery ventilation is a more recent technology for improving indoor air quality. Heat recovery ventilators work by filtering and recirculating air through the building, using a heat exchanger to capture the heat from outgoing, stale air and using it to warm incoming, fresh air. It can be connected to several rooms with a duct system. These systems also help to control humidity. Heat recovery ventilators are used extensively in Europe and North America as a way of retaining energy within the building and providing a healthier living environment. This may be a good option for people who live in cold climates, particularly those who have respiratory sensitivities.

Top DIY products for cleaning the air
- **The garden shed** is the best place to store garden chemicals, paints, solvents, cleaning products and other chemicals.
- **Exhaust fans** remove mould-causing moisture from bathrooms and kitchens, combustion gases from cooktops, and fumes and pollutants from workshop areas.
- **Whirligigs** ventilate the roof space of houses. They are also a great option to improve garage, shed or workshop ventilation.
- **Flyscreens** let the fresh air in, but not the flies and mozzies.

Air filters

Where indoor air is referred to as 'stale', outside air is often referred to as 'fresh' or 'clean'. But what if you live in an area with high air pollution? In such cases you may wish to invest in an air cleaner or filter. These devices are designed to remove pollutants from the air in a variety of ways. They might use HEPA (high efficiency particulate arrester) filters to remove fine particulate matter, ion generators to remove airborne particles using an electronic charge, and/or activated charcoal filters to remove some gases. Keep in mind that these systems can be expensive, and they need to be maintained and their filter media replaced regularly in order to remain effective.

Ventilation as part of a 'whole house'

Many ventilation systems greatly affect the performance of combustion appliances such as furnaces, water heaters and fireplaces, which generally have their own exhaust systems. If you live in a cold climate and need mechanical ventilation, it is well worth talking to a home energy specialist and asking their advice for your particular needs and home circumstances. Ventilation, space heating and cooling, water heating, cooking, clothes drying and dehumidifying are often all interrelated. They can share heat sources, ducts, chimneys and exhaust fans. Optimal efficiency and performance can be achieved if you take a whole-house approach to choosing and using appliances that perform these functions.

Overview of common pollutants and what to do about them

Source*	POTENTIAL POLLUTANTS*	TIPS TO REDUCE POLLUTANTS
Occupant activities and lifestyle		
Smoking	Particulates, benzene, polycyclic aromatic hydrocarbons (PAHs)	• Don't smoke indoors.
Petrol and car exhaust vapours from attached garages	Benzene, carbon monoxide, VOCs	• Ventilate garage.
Fuel-based heating and cooking, especially if unflued	Butane, carbon monoxide, limonene, n-hexane, propane, particulates, acrolein, nitrogen oxides, formaldehyde	• Use range hood fan when cooking. Make sure it expels air outdoors, not back into the room. • Get rid of unflued heaters. • Have heaters serviced and flue checked every couple of years.
Construction and DIY products (dry)		
Plastic and rubber flooring and carpet underlay	VOCs	• Choose flooring alternatives. • Ventilate regularly.
Carpet	Dust mites, 4-vinylcyclohexene	• Consider sealed floors instead. • Vacuum regularly with a vacuum cleaner with high-performance filters (e.g. HEPA)

Source*	POTENTIAL POLLUTANTS*	TIPS TO REDUCE POLLUTANTS
Wallpapers	Mould inhibitors	Choose with care
Construction and DIY products (wet)		
Adhesives and sealants	Acetone, ethanol, formaldehyde, methanol, VOCs	• Choose products that are environmentally certified (e.g. Good Environmental Choice or Sensitive Choice certified), or at the very least, those that are low-fume or low-VOC. • Ventilate during use and as frequently as possible in the months after use.
Timber stains, paints, coatings	Ethanol, methanol, methyl chloroform, xylene, toluene, VOCs, lead	
Polyurethane lacquer/ floor varnish, including those used on concrete	Acetone, benzene, xylene, toluene, isobutyraldehyde	
Building contents		
Furniture	Dust mites, VOCs, formaldehyde	• Consider second-hand furniture, which typically has already off-gassed most fumes and pollutants • Favour solid timber over particle or composite board, or look for composites made with low-fume adhesives. • Regularly vacuum furniture to control dust mite allergens.
Office equipment	Ozone, VOCs, particulates	• Regularly ventilate the rooms where equipment is used, particularly in the first few months after purchase.
Cupboards and shelving	Formaldehyde, nonanal	• Favour solid timber over particle or composite board, or look for composites made with low-fume adhesives.
Dry-cleaned clothing	Methyl chloroform, tetrachloroethylene	• Look for a greener dry-cleaner that uses a non-toxic alternative, such as a 'wet cleaning' process or carbon dioxide cleaning. • With conventionally dry-cleaned clothes, remove the plastic and air them outside before returning them to the wardrobe.
Printed material	Formaldehyde, nonanal, toluene	
Waxes and polishes	VOCs	• Choose environmentally certified, low-fume and/or low-VOC products.
Cleaners, disinfectants and detergents	Benzene, butane, ethanol, toluene, formaldehyde, limonene, methanol	• Choose environmentally certified, low-fume and/or low-VOC products. • Consider simple green cleaners, such as bicarb soda, instead. • Ventilate during use.

Source*	POTENTIAL POLLUTANTS*	TIPS TO REDUCE POLLUTANTS
Cosmetics	Ethanol, methanol, nonanal	• Look for greener alternative brands. Read ingredient lists. • Avoid aerosol-dispensed products. • At the very least, apply fumy products in the bathroom with the exhaust fan running.
Room deodorisers	p-dichlorobenzene	• Avoid using them—remove cause of odours instead.
Biological sources		
Mould, spores and fungi	Allergens	• Address moisture problems that cause mould.
Garden plants	Pollen and other allergens	• If a problem, consider changing to an asthma-friendly garden—see www.asthmafoundation.org.au.
Pets	Allergens, such as pet dander, saliva, faeces and urine	• Keep pets, pet areas and bedding clean. • Groom pets outdoors.

* Sources and pollutants adapted from Environment Australia, *Air Toxics and Indoor Air Quality in Australia*, Department of Sustainability, Environment, Water, Population and Communities, Canberra, 2001.

Removing excess moisture to control humidity

Air that is too dry can make our skin dry and itchy, dry out and irritate our nasal passages, and make it harder to breathe. When you consider that our bodies are around 65% water—blood is 82% water and the lungs are 90% water—it's no surprise that our bodies need some moisture in the air to ensure our comfort. However, too much moisture is also a problem. Air that is too damp can also irritate skin and nasal passages. Humid indoor air can also cause condensation on windows and water damage to soft furnishings, promote the growth of mould and even rot wood.

The easiest way to control excess moisture and high humidity is to prevent the build-up of moisture in the first place.

Tips for excess moisture prevention

• First check whether you have adequate ventilation. If you don't have a mechanical ventilation system and you suffer from asthma or allergies, chances are that your health and moisture problems may both be eased by investing in an energy-efficient ventilator.
• Use a bathroom fan to remove steam moisture from baths and showers. A good-quality, energy-efficient model will do this while reducing heat loss from the house.
• Air your house on dry, hot days rather than in muggy weather.
• Check for leaking pipes and water fixtures. Repair leaks promptly.

- Have clothes dryers vented to the outside air if indoor moisture is a problem.
- Poor site drainage on your property can contribute to moisture problems. Grade the soil so that water naturally flows away from your house, aided by gravity.
- Use a fan when cooking to remove steam as well as combustion gases.

Humidity can be controlled to keep the relative humidity inside your house at an ideal level of below 50% in summer and 30% in winter. You can check your home's humidity with a hygrometer, which costs around $20–100. If your best efforts aren't keeping moisture under control, then you may need a dehumidifier.

Prices for dehumidifiers range from under $100 for a small portable room dehumidifier up to several hundred dollars for a good-quality, energy-efficient model suitable for a large house with major moisture problems. A dehumidifier is an energy-using appliance, so it's wise to factor in the running costs as well as the purchase price. As with other energy-using appliances, reduced energy costs generally mean reduced environmental impact.

Conversely, there may be times when the air is too dry for comfort. Humidifiers (or steam vaporisers) are commonly available for around $50. They can be bought from chemists and department stores. They are often used to ease breathing congestion in people with colds or asthma and can be used in conjunction with inhalants, such as eucalyptus oil, that help to clear the nasal passages. If you use eucalyptus oil with a humidifier, remember always to keep both the oil and the humidifier well out of reach of children, as eucalyptus oil is poisonous if swallowed. Humidifiers can also pose a scalding risk, so always follow product instructions.

→ **ASK TANYA**

My housemates will finish showering and leave the bathroom extractor fan plus the overhead fluorescent light (these are connected to the same switch) on for *at least* half an hour after they've left the room. I, on the other hand, will turn everything off but leave the bathroom door open.

Can you please tell me how much electricity an extractor fan + fluorescent bulb are likely to use, whether the fan or door method is likely to be better at 'de-fogging' a bathroom, and if there's any good reason to not just go with the open door method?

Sarah

A compact fluorescent bathroom lamp will use up to 20 watts, and bathroom fans typically use 40 watts. (Note for other readers: heat lamps use much more: up to 375 watts per lamp.) Together, they'll use 0.03 kilowatt hours of electricity over half an hour—not a huge amount, but every bit adds up. Whether an open door or fan is better at defogging depends very much on where you live and the season. In a dry Adelaide summer, I'd leave the door (and a window) open, but during the wet season in Australia's tropical north, an extractor fan would be more reliable than natural evaporation as the air is closer to saturation and consequently there's greater risk of moisture problems that rot wood and encourage mould. Mould can make your living environment very unhealthy; I'm not just being a neat freak!

ENDOCRINE DISRUPTORS—CHEMICALS THAT MESS WITH YOUR HORMONES

There are some chemicals that do a very good impersonation of hormones. These substances are called endocrine disruptors because there is evidence to suggest that they can influence and damage the endocrine systems of the animals and humans that ingest them.

The human body's endocrine system includes glands such as the pituitary, adrenals, ovaries and testes. It also includes the hormones that they produce—little chemical messengers sent through the body via the bloodstream.

Endocrine disruptors can reproduce the effects of hormones, most commonly the female hormone oestrogen, when they enter the body. Too much oestrogen can reduce fertility, harm foetal development and produce female physical characteristics in males. However, it appears that small amounts of a single endocrine disruptor have little effect.

The cause for concern with endocrine-disrupting plastics is that there are so many of them and they surround us. Researchers at Tulane University in New Orleans reported that when a number of weak endocrine-disrupting chemicals were combined, they became many times more potent. Endocrine disruptors are also fat-soluble, so they can accumulate in the food chain, and have a long biological half-life.

Endocrine disruptors are virtually impossible to avoid completely because they're so prevalent, but you can take measures to limit your exposure to them.

Basic types of endocrine-disrupting substances

- **Pesticides**, such as DDT (excluding natural alternatives such as pyrethrum)
- **Heavy metals**, such as lead, mercury and cadmium
- **Organochlorines**, which are compounds made from carbon, hydrogen and chlorine; examples include dioxins (often associated with chlorine bleaching), PVC and polychlorinated biphenyls (PCBs)
- **Bisphenol A (BPA)**, which is used to make polycarbonate plastic and epoxy resins
- **Plasticisers and surfactants**, which are often used to make plastics; examples include phthalates (such as polyethylene terephthalate, which is used to make soft-drink bottle plastic), polycarbonates and styrenes

Tips to avoid exposure to endocrine disruptors

- Fats tend to absorb endocrine disruptors, so to avoid the migration of these substances into food, avoid storing fatty food in plastic containers or plastic wrap.

- Don't heat food, particularly fatty food, in plastic containers.
- Avoid oily fish and particularly shellfish harvested from polluted waters. Shellfish tend to accumulate heavy metals. Fortunately, our oceans are relatively unpolluted, so Australians are reasonably safe eating local seafood.
- Avoid pesticides in the food you prepare by buying organic food where possible. Thoroughly wash or peel fruit and vegetables.
- Avoid the use of pesticides in your home and garden.
- Only use plastic water bottles that state they are BPA-free.
- Ask if pesticides are used at your workplace or your child's school. If they do use them, politely suggest some non-toxic alternatives. For more information on these alternatives, see chapter 4, pages 55–9 and chapter 11, pages 213–16.
- Don't give children (particularly teething tots) soft plastic toys or teething rings unless they state that they are free from phthalate plasticisers. Avoid toys made from PVC, especially if your child tends to put toys into his or her mouth.

MAKE A COMMITMENT

This week I will...

☐ _____

☐ _____

☐ _____

This month I will...

☐ _____

☐ _____

☐ _____

When I get the opportunity I will...

☐ _____

☐ _____

☐ _____

Notes

_____

_____

_____

_____

_____

3

Working green

Just as you can make changes to your lifestyle so that you tread more lightly on the earth at home, you can also change the way you work. It's even easier to make a difference if you have an office at home, where you are in control.

If you are employed outside the home, don't leave your values on the doorstep when you leave for work; while you may not be in a decision-making position, you can certainly contribute to a healthier workplace and healthier planet by looking at your own working habits and making changes to the way you do things. If you are in management, reducing costs, improving energy efficiency, building your organisation's reputation and looking after worker wellbeing may well be a core part of your job.

There are so many different kinds of workplace that it's hard to make environmentally helpful suggestions that apply across the board, but there are a few common elements, most of which relate to administrative tasks and energy consumption.

All businesses include some form of administration and paperwork, and they all use resources, produce waste and ultimately affect the environment. Many of us have made attempts at creating a paperless office and have looked into recycling office wastepaper and buying copy paper with recycled content, but there are other green issues in the average office, including energy use, water use and the core business practices themselves. And keep in mind that we're now living in an era where 'green is the new black'— greening up your workplace is an opportunity.

WHY GO GREEN AT WORK?
While many CEOs make grand and caring, sharing statements about corporate responsibility, it's not all about bleeding-heart altruism. Just as anyone can benefit from the money savings, comfort and healthier environment of a greener lifestyle, all workplaces can benefit from improving their sustainability practices. Ultimately, going green simply makes good business sense.

Reduced running costs

Reducing the consumption of energy, water and material goods used in the office will save money. This may involve simple changes like reducing over-illumination, cutting back paper use or using energy-saving features on computer equipment. For those businesses that pay for waste-management services, reducing waste can also save money.

Reputation benefits

Solid environmental policies that are well implemented are something you can brag about. This may be a selling feature for your product or services, or it may even allow some companies to be included in ethical investment funds. Many companies are now including the measurement of their green-house emissions in their annual reporting.

A good reputation can help a company attract and retain good staff. Corporate social responsibility and reputation are central to what human resources experts call 'pride affiliation'—people want to feel proud, not embarrassed, to say who they work for, making companies with good green credentials the employers of choice for workers with a social conscience.

A healthier, more comfortable workplace

Improving the energy efficiency and thermal comfort of an office makes it a more comfortable place to work in. You can also choose cleaning products, furnishings and other items with a view to improving indoor air quality, which also looks after the health of the people you work with.

Do workers and consumers want green?

A 2007 survey of 16,823 consumers in 15 countries found that 53% of consumers prefer to purchase from companies with a strong environmental reputation. The research conducted by Ipsos MORI also found that eight in ten workers would prefer to work for an environmentally ethical organisation.

Good morale and increased productivity

A healthier, more comfortable working environment means happier workers. It can also result in significant decreases in absenteeism, headaches, colds and flu, fatigue, poor concentration and other worker health and well-being concerns. These improvements are linked with improved productivity, which is great for business, particularly for those organisations whose largest costs are employee wages.

ENERGY

Getting down to business, a good place to start is by focusing on your workplace energy consumption. The energy tips included here apply well to home offices and small businesses. For larger organisations, it's worth having an energy audit conducted on your business or joining one of the many government-run business sustainability programs. A trained and qualified energy auditor can look at the unique features of your office space, equipment and systems and assess where improvements can be made.

So what are the big office energy users? The pie chart pictured here shows where energy is used in offices at the big end of town. Heating, cooling and ventilation together are the largest energy demand for Australian commercial buildings, followed by lighting and office equipment. For the humble home office, running electronic equipment becomes more significant. Here are some ways to improve energy efficiency in these three big areas of office energy use.

Commercial building energy use

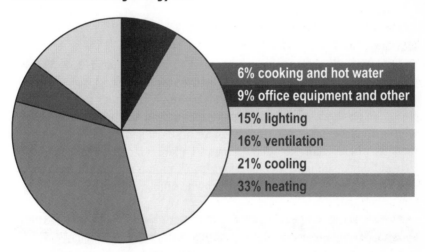

6% cooking and hot water
9% office equipment and other
15% lighting
16% ventilation
21% cooling
33% heating

Source: *Australian Commercial Building Sector Greenhouse Gas Emissions 1990–2010*, Australian Greenhouse Office, Canberra, 1999

Heating, cooling and ventilation

In building-management circles, heating, ventilation and airconditioning are collectively known by the acronym 'HVAC'. For home offices, the heating and cooling advice in chapter 9, pages 127–44, and the indoor air quality advice on pages 11–19 apply. For larger corporate offices and places of business, the following tips apply:

- Airconditioning and heating systems should be zoned, so that separate work areas or floors can be separately controlled, rather than heating or cooling the whole building.
- As for home, set thermostats to a reasonable temperature of 18–20°C in winter and 23–26°C in summer.
- Make sure your HVAC systems are regularly serviced and maintained. Well-maintained HVAC systems tend to be more efficient, more effective and cheaper to run, and provide a more comfortable and healthier working environment.
- Use timer controls so that HVAC systems are used only when the building is occupied. They can be programmed to turn on 30 minutes before the working day begins, and to turn off at the end of the day.
- Bear in mind that lighting can have an effect on HVAC needs. For example, a business that has halogen downlights, chosen for their good looks, fitted in its reception area may find that the waste heat from these lights makes this area particularly hot, especially in summer. Swapping to LEDs will reduce both lighting costs and cooling needs.
- Make sure the building has enough shade and insulation to reduce the need for heating and cooling.

Lighting

The same home lighting principles covered in chapter 9 (pages 157–63) also apply to the home office or study. If you can, choose office space with large windows. Natural lighting will lift your spirits and, what's more, it's free. North-facing windows will allow good solar access, and provide free heat in winter. However, you may find the glare of direct sunlight a problem, particularly when placing computer screens. If glare is a problem and you want to avoid direct sunlight, then consider using a room with south-facing windows for a home office and this will provide diffuse natural light.

AREA	LIGHT NEEDED	RECOMMENDED LIGHTING LEVEL (LUX)
Corridors, walkways, storage rooms, change rooms	Enough to allow movement and orientation	40–100
Waiting rooms, lunchrooms, common areas	Background lighting, and for simple tasks and short periods of reading clear text	160
Desks, work benches, food preparation areas	Task lighting and for general reading	300–400
Some desks and work spaces	Enough for visually demanding tasks, such as proofreading or fine machine work	600

For a commercial office space, different areas—such as desks, washrooms, reception areas and stairwells—have different light needs. It's important to provide light that's 'fit for purpose'; it's a false economy to read in poor light to save energy only to ruin your eyesight. The table on page 26 is a guide to lighting needs in a typical workplace.

Lighting tips

- Aim for a lighting system with a total energy use of 5–8 watts per square metre of floor space. As LEDs become cheaper and more commonly available, you might be able to get this down to as little as 2 watts per square metre.
- If you haven't already, swap incandescent light bulbs for compact fluorescent or LED alternatives.
- Don't forget to include lamps in your cleaning and maintenance plans, as dust can significantly reduce light output.
- The more control you have over lighting, the greater the potential to improve its efficiency. For new systems, put in separate controls for the lighting of separate areas, and install multiple switches instead of a single switch to control a large number of lights. Consider dimmer controls, particularly in areas that get variable amounts of natural light. Make sure the switches are clearly labelled.
- Consider motion-sensor controls for lighting areas that are only occasionally used, such as washrooms or storerooms. Also use motion-sensor controls for outdoor security lighting.
- Install reflectors so that you get the most out of your lighting.
- For new fluorescent tube fittings, choose the new T5 (16 mm diameter) tubes and fittings, which are more energy-efficient than the older T12 (38 mm diameter) and T8 (26 mm diameter) tubes. T5 tubes use about 25% less electricity than T8 tubes, produce more light and contain less mercury. Note that T8 and T12 tubes fit into G13 base connections, while T5 tubes are slightly shorter and require different base connections, so get the right base from the outset.
- Depending on your existing fitting, you may be able to get retrofit kits that adapt T12 and T8 fluorescent tube fittings to take the more efficient T5 tubes. These kits offer the option of reflective or white backings.

De-lux!

Many homes and offices are lit more than they need to be. Aside from the 'headache' of large electricity bills, this over-illumination is linked with health issues such as headaches, higher levels of stress and worker fatigue. Lux meters, devices that measure lighting levels, are useful for checking lighting levels and are a tool of trade for energy assessors. They cost around $50 to buy, but a handful of public libraries and local council sustainability departments have them available for short-term loan. Or you might be able to borrow one from a photographer friend.

- Retrofitting with reflectors and brighter, more efficient T5 tubes may allow you to remove excess lighting fixtures (also known as 'de-lamping').
 - Don't forget to recycle old fluorescent tubes and compact lamps. See page 163.

Appliances and devices

Office equipment is another big electricity user in the workplace. With an ever-growing range of computers and devices we can plug in and put to work, the power needs of office equipment threaten to gobble up the gains we make by saving energy elsewhere. However, this is one area where some simple changes can make a big difference.

Computers

- Monitors account for most of the power used by your computer when it is on. When you shut down your computer, make sure the monitor is off as well.
- Don't use screen savers because they often use more electricity. Instead, use the screen and hard-drive sleep modes when your computer is left unattended, which will save energy and protect the screen.
- Laptops can use up to 90% less energy than desktop computers. Depending on your budget and the way you use computers, consider a laptop instead of a desktop computer.
- Internal devices, such as internal modems or CD-ROM drives, are generally more efficient than external devices because internal devices run off your computer's own power supply. External devices generally have separate power connections.
- Remember when buying a monitor that the larger the screen is, the more energy it will consume.
- When buying office equipment and computers, look for the ENERGY STAR symbol, which shows that the device has met certain standards for energy efficiency.
- Make sure the power-management and energy-saving features of your computer are turned on. These are usually controlled through the system preferences of your computer.
- Even electronic documents need to be filed and disposed of if not needed. Filling your computer's hard drive with large, memory-using files can slow down the computer, reducing your productivity.
- Remember to turn the printer off rather than leave it on standby.
- Inkjet printers use up to 95% less energy than laser printers.

- Share printers in an office environment, rather than having one for each computer.
- Save paper by using print preview functions to see whether you need to change some of the page set-up or layout.

Photocopiers

- Photocopiers use huge amounts of energy, so choose the smallest size that will meet your needs.
- Choose photocopiers that can print double-sided.
- Check that the photocopier you buy can use recycled-content copy paper.
- Photocopiers run more efficiently when they print large amounts, so do photocopying in batches. For a small number of copies use your printer instead.

Corporate energy use

Australia's commercial sector used 268 petajoules (or 10^{15} joules) in the financial year 2007–08, a 22% increase from the 219 petajoules used in 1999–2000.

Facsimile machines and multifunction devices

- Consider buying a printer and fax machine in one, particularly for small offices or home offices. There are also 'multifunction devices' that act as scanners in addition to the other functions. They use less electricity than the combined total of several separate devices, and they take up less space.
- Fax-modems are also available, which allow you to send electronic documents as faxes directly without having to print them first.

Standby power

Standby power demands, also known as 'phantom loads', are problems in most parts of the house, and a home office is no exception. Remember that some electrical appliances connected to the power point still draw small but continuous amounts of electricity when they're in standby mode and even when they're turned off.

Some appliances need to stay connected and turned on 24 hours a day. These include the kitchen refrigerator, fax machines, telephone systems and answering machines, and some kinds of photocopier (those that have a small electric element to prevent moisture from building up inside). However, others can be turned off at the wall at night and over weekends to reduce the costs of standby power.

If you want to turn certain appliances off at the power point at night to prevent energy wastage, mark the different power cords or points with coloured dot stickers. Use a different colour for the appliances that should be left on so that you don't confuse them. Alternatively, group them on two separate power boards. If some power points are hard to get at, consider plugging them into a more accessible power board. Plug-in timers can also be put into power points. For more information on standby power, see page 165.

For more information on standby power, see page 165.

→ **ASK TANYA**

In a work environment with 100 staff, what would be the greener option: burning coal to operate several electric hand driers or the environmental damage of the waste accumulated by countless paper towels (even if we do choose the recycled ones)?

Fiona, Fortitude Valley, Queensland

Along with cloth versus disposable nappies, and beer in cans versus stubbies, this is one of those great conundrums. Hand dryers obviously come out in front when comparing material use, waste generation, and ongoing costs and upkeep. Less obvious is the energy consideration. Yes, dryers use electricity, but paper towels also consume energy and water in growing the raw materials, refining them, manufacturing the towel, transporting it, and collecting and disposing of the waste.

I'm yet to see an Australian life-cycle study of this conundrum—undoubtedly our scientists have their hands full with more pressing issues—but overseas articles say that hand dryers represent less total energy use. Newer, more energy-efficient hand dryers are coming on the market, which use a shorter (10 seconds instead of 30–40 seconds) but more intense burst of air.

So my view is choose the hand dryer option and run it with electricity from cleaner, renewable GreenPower-accredited sources. Burning coal isn't the only option for your workplace's electricity needs. Visit www.greenpower.gov.au for more information.

WATER

- Install water-efficient taps and flow restrictors in washrooms.
- Encourage staff to report dripping taps or leaking toilet cisterns and have them fixed promptly.
- Change over single-flush toilets to water-efficient dual-flush toilets. In some areas there may be government rebates available to assist businesses to do so. Contact your water authority and see what's available.
- Put some signs up! The same home water-saving tips apply at work, but it may be harder to educate colleagues than to educate housemates. Put a note on the staffroom dishwasher reminding colleagues to use it for full loads only, put some energy-saving tips on the back of toilet cubicle doors (where you have a captive audience), and so on.
- Make sure any landscaping is water-wise. Use mulch on garden beds and timer systems for watering if it's needed.

→ **ASK TANYA**

I work in a shopping centre and was wondering, when the call of nature comes, what uses less water? In one of our toilets, we have a stainless steel wall urinal as well as normal half/full flush toilets. In another toilet area the shopping centre has installed what look like more water-efficient single-user ceramic urinals. Does anyone have any idea, especially with the wall-type urinals, what uses less water? I apologise if this is a silly subject, but it concerns me.

Richard

CONSUMABLES AND WASTE

Office consumables include copy paper, toilet tissue, assorted stationery, light globes, toner and ink cartridges, CDs and other products. All consume energy, raw materials and water in their manufacture. Some use these resources unsustainably. These products also end up as waste.

Paper: reduce, reuse and recycle

Offices generate waste through the use of paper, other stationery products and goods with packaging.

For those who control the office purse strings, paper represents a big cost. To the more green-minded, paper production uses a lot of energy and water, has often been bleached, pollutes waterways, and is often based on virgin fibre from woodchips. The more paper an office uses, the higher both the dollar costs and the environmental costs.

Paper by numbers

In 2008–09, the total Australian consumption of printing and writing paper was 1.733 million tonnes. The Organisation for Economic Co-operation and Development (OECD) predicts that global paper consumption will have grown by around 77% by 2020.

To cut down these costs:

- When given a choice, subscribe to the electronic versions of newsletters, rather than the printed versions.
- Get your printing estimates right. Printing extra copies of literature only wastes paper, costs more money and takes up more storage space.
- Proofread and spellcheck your documents on screen. As a rule, make sure you at least run a spellcheck before printing a document. Use the 'track changes' function in Word programs to make editing notes.
- See if you can make all your fax communication digital.
- Set your printer and photocopier to print double-sided.

- Reuse paper that's only printed on one side as notepaper or for printing draft documents. Keep a separate tray or bin for it next to your printer or photocopier.
- Reuse envelopes if presentation-quality envelopes aren't needed.
- Recycle paper that has been printed on both sides. Also recycle envelopes, newspapers, magazines and other paper products. Do not include self-carbonating paper as it contaminates the recycled product.

Once we've exhausted the potential to reduce paper consumption and reuse paper where possible, the next step is recycling. There are now many established businesses and a market for recycling post-consumer office paper. Paper and cardboard represent about 70% of all the waste produced by offices. Many paper-recycling companies comment that it is difficult to get reliable clean supplies of recycled feedstock, but that the quality of paper from offices and manufacturers is much higher than from households. Companies like Australian Paper and Visy provide collection services for white office paper, which is then recycled in Australia. Some also collect cardboard and newspapers.

Helping hands for saving paper

These two initiatives have case studies, tips, information and software to help businesses reduce their paper consumption:
- Do Something! Save Paper at www.paperlessalliance.com.au
- PaperCutz at papercutz.planetark.org.

Larger businesses are probably already paying waste-disposal costs. Such companies can have recycling built into their waste-management contract. In addition, if you first reduce the amount of waste your organisation produces you can dramatically cut your waste-disposal costs.

E-waste recycling

Paper isn't the only material in a business environment that can be recycled.

Mobile phones and their **batteries** should never be sent to landfill. Mobile phone batteries can contain the heavy metal cadmium, which can leach from tips into groundwater and threaten environmental health. The good news is that mobile phones and their batteries can now be reused or recycled in Australia through a number of collection programs. The Aussie

Recycling Program (www.arp.net.au) collects unwanted phones, refurbishes them and sends them to developing countries where people generally can't afford new phones. Some charities, such as Clean Up Australia and Flora and Fauna International (www.fauna-flora.org.au), collect on their behalf and receive a donation per handset collected. These collection schemes have various drop-off locations, or they can send out reply-paid satchels to people with unwanted phones.

The Australian Mobile Telecommunications Association's 'Mobile Muster' program collects mobile phones and recycles them, recovering their various composite materials. Handsets are collected through recycling bins in many retail mobile phone stores or via replay-paid satchels. Find store locations and reply-paid instructions at www.mobilemuster.com.au.

Printer cartridges (laser and inkjet) and **toner** bottles can be refilled, re-inked, remanufactured or recycled, with widely varying environmental benefits. They should not be sent to landfill as this both wastes the resources that were used to make them and contributes to pollution. Toner and ink can contain cadmium, carbon black and a range of other potentially hazardous materials.

Many companies will offer to take back (and occasionally even pay for) some models of empty cartridge. While there are literally hundreds of different types and brands of cartridge, very few are suited to remanufacturing, the process in which a cartridge is opened, various components are replaced and the cartridge is refilled with toner. In many cases, remanufacturing operations will remanufacture certain types of cartridge and dump the remainder, which includes most inkjet cartridges, in landfill.

The 'Cartridges 4 Planet Ark' program offers recycling for all types of empty printer cartridge. Planet Ark has set up special printer cartridge recycling bins in retail outlets around the country, including participating Australia Post and Officeworks stores. The public can drop off any printer cartridge at these points for recycling—regardless of its make or condition. Organisations that use large numbers of cartridges can apply to have their own free 'Cartridges 4 Planet Ark' collection service, with their own bin within their office. Over 50% of the laser cartridges collected are sent to the original equipment manufacturers for their remanufacturing or component recovery programs. All remaining cartridges, toner bottles and drum units are processed to recycle their component materials into new products. To find your nearest cartridge recycling retail outlet or to apply for the business collection program, visit cartridges.planetark.org.

E-waste facts and figures

Electronic waste or e-waste is a nasty side effect of our love affair with new technology.

- Every year 20–50 million tonnes of e-waste are generated around the world.
- In 2007–08, 31.7 million new televisions, computers and computer products were sold in Australia, while 16.8 million (106,000 tonnes of e-waste) reached the end of their life.
- In 1997 the average life span of a computer was 4–6 years. By 2005 this had dropped to two years.
- E-waste can contain a number of hazardous materials including lead, cadmium, mercury, chromium VI and brominated flame-retardants.

Given the amount of energy and material that goes into making cartridges and the potential for pollution with dumped cartridges, it is best to reduce your use of them by reducing the amount of printing you do. This will also save paper.

The used cartridge trade is a lucrative business and there are a handful of 'backyard' remanufacturers producing poor-quality cartridges—more interested in a quick dollar than the environment, whatever their sales pitch may be. Poorly remanufactured cartridges can cause extra wear and tear on the printer and shorten its life span. Make sure that any remanufactured product you buy comes with a guarantee and is remanufactured in a facility that is ISO 14000 certified.

Computers and **computer parts** can be reused, refurbished or recycled into new computer parts, or the valuable metals and other materials can be recovered for other purposes. It's advisable that you first delete all your personal data and files. There are a number of government, charitable and commercial computer recycling programs that take unwanted computers from businesses and home users. To find out where you can recycle computers and other office-related items, visit Planet Ark's Business Recycling website at businessrecycling.com.au.

As part of its product stewardship activities, IBM offers its clients take-back services on obsolete equipment, and has a buy-back program for anything that may be of commercial value. The IBM Certified Used Equipment program sells certified refurbished second-hand PCs at a fraction of the cost of new models and also donates some equipment to charities.

Never put computers or unwanted electronics in with general rubbish as they contain a number of potentially hazardous materials. E-waste from discarded technology is a growing and serious environmental problem.

Other office waste reduction and recycling

The lunchroom is another place where waste can be recycled. Set up recycling bins for drink **bottles and cans**, as you would at home. For apple cores and other food scraps, self-contained indoor composting bins called Bokashi bins are well suited to office food waste processing. See page 106 for more on Bokashi bins.

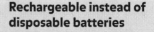

Rechargeable instead of disposable batteries

With wireless mice, notetakers, digital cameras and many other gadgets needing batteries, you can now buy AA batteries in packs of 40. Consider buying rechargeable batteries and a recharging unit instead of relying on disposable batteries. This will cut waste and save money in the long run. Based on an average saving of $1.50 per battery, a $40 recharging system will pay for itself in the time it takes to use 27 batteries' worth of energy.

Choose flash drives or rewritable **compact discs (CDs)** and **DVDs** for saving files that you plan to update, change or add to, instead of using single-use discs, then throwing them out when they're no longer needed. Use single-use discs for archiving files, as these discs are better suited to long-term file storage than rewritable discs. There are limited recycling services available for unwanted discs. Collection points are included on Planet Ark's Business Recycling website at businessrecycling.com.au. Note that the details of each collection organisation will include a list of the materials they collect. If CDs and DVDs are listed followed by the word 'Recycle' in brackets, then all types of disc are accepted for recycling to recover their component materials. If the word 'Reuse' follows, this means that the organisation is probably a charity collection looking for music CDs and movie DVDs to resell, as opposed to data discs.

Office furniture can be sold second-hand, refurbished or recycled.

PURCHASING

Green purchasing aims to make a company truly 'eco', by balancing the considerations of both the ecology and the economy.

Organisations such as ECO-Buy and the Buy Recycled Business Alliance (BRBA) are encouraging businesses to put their purchasing dollar behind greener products and services. ECO-Buy offers information, advice and other support to members in both the business and government sectors nationally. In late 2005, Toyota Australia became the first ECO-Buy Business member. Any business interested in helping the environment by purchasing greener supplies should have

> **Can coffee-to-go go green?**
>
> Polystyrene foam cups are not biodegradable and are rarely recycled. And even though paper cups come from trees, they're still a single-use disposable item. Take your own mug to get takeaway coffee or tea. Alternatively, try the KeepCup—a clever little Australian-made, reusable takeaway hot-drink cup. It's made from BPA-free plastic, is insulated and is sized to be barista-friendly, fitting comfortably under the spouts of espresso machines. Visit: www.keepcup.com.

 a look at ECO-Buy. For more information, visit www.ecobuy.org.au. The ECO-Buy website also includes the ECO-Find searchable database of green products.

No matter how small your business is, you can adopt a purchasing policy that favours energy-efficient products, products with lower chemical content, greenhouse-friendly products and products made from alternative green fibres, as well as recycled products. The reduced energy costs from buying energy-rated products and energy-efficient light bulbs will save quite a bit of money. When it comes to energy, being green can keep you out of the red. Consider reinvesting these savings in the environment, by using the money to subsidise the sometimes higher cost of environmentally preferred cleaning products and recycled-content papers.

Paper products

The paper you choose is as important as how you use it. The two main considerations are the fibre content and the way it is (or isn't) bleached. Ideally, aim to purchase paper made from recycled waste or sustainable alternative fibres, such as hemp or sugarcane waste. There's nothing wrong with paper made from wood pulp, as long as the wood comes from sustainably managed timber plantations, not old-growth forests. There does need to be some virgin fibre used at some stage, as paper cannot be infinitely recycled. Eventually the fibre disintegrates. However, paper does not need to contain 100% new, high-quality fibre. The bulk of timber from sustainable sources should primarily be for use in building and furniture, where the structural strength of hardwoods is needed.

Look out for the following when purchasing office paper:

- Avoid papers that have been bleached with chlorine or chlorine compounds, which produce toxic dioxin compounds as a by-product. Those that are labelled 'Elemental Chlorine Free' or ECF don't use molecular chlorine gas as the bleach, but typically use chlorine dioxide, which still produce dioxins, though to a lesser extent than elemental chlorine. Instead use varieties that are oxygen-bleached or unbleached or that state that they are 'Total Chlorine Free' or TCF bleached. Some may be 'Process Chlorine Free' or PCF, which means that they may have recycled content that was previously bleached, but the manufacturer hasn't used any further chlorine-containing bleach.
- Look for office papers with a high percentage of recycled content that comes from 'post-consumer' waste—wastepaper that has been used at least once by consumers, after which it is collected and sorted. Pre-consumer waste is often offcuts from the production of other virgin papers or sawmill waste, which is essentially virgin pulp. In Australia, some of this comes from old-growth forests.
- Look for paper products that are Forest Stewardship Council (FSC) certified as coming from sustainably managed forests.
- Visit the ECO-Buy website at www.ecobuy.org.au to find bulk suppliers for recycled-content office paper.
- The Sustainable Choice program of the New South Wales Local Government Association has a useful fact sheet on paper choices for copy paper available in New South Wales. It can be viewed online in the 'Facts and Figures' section at www.lgsa-plus.net.au/sustainablechoice.

- Avoid papers made from plantation fibre from unknown sources. They're likely to come from poorly managed plantations in Indonesia, Thailand or Brazil.

→ **ASK TANYA**

I've seen various logos on reams of copy paper. Which ones are the good ones?

Joseph, Moorleah, Tasmania

As a rule of thumb, look for the logos of independent, third-party certifiers. Good Environmental Choice Australia (GECA) and the Forest Stewardship Council (FSC) are both independent environmental certification organisations operating in Australia. There are also programs similar to GECA overseas, such as Green Seal (USA), EcoLogo (Canada), Nordic Swan (Scandinavia) and Blue Angel (Germany). Products certified by these programs may be sold in Australia.

Take private eco-labels or logos (those owned by a brand and/or not open to use by their competitors) with a grain of salt and always seek further information. A logo with a leaf or a happy-looking tree is meaningless without clarification of exactly how a product is a better environmental choice.

Aside from recycled paper, what are the other types of paper that aren't made from trees?

Sky, Attadale, WA

The Environmental Paper Network's State of the Paper Industry report estimates that about 9% of paper fibre comes from non-wood sources globally. The main non-wood fibres are straw, sugarcane bagasse and bamboo. Other fibre sources include cotton, reeds, sisal, hemp, jute, kenaf, flax and banana (using the stem and any non-edible fruit). The fibres that come from agricultural residues are promising, particularly in cases where they would otherwise be burnt off, contributing to air pollution. But, for novelty value, you can't beat panda poo paper, made from the fibre-rich excrement of giant pandas living in captivity. But you might have trouble getting it through customs.

GreenPower

The money saved through office energy-efficiency programs can be put towards purchasing GreenPower. For further information on GreenPower, see page 169. At first glance it might seem crazy to spend more money for electricity than you have to. However, choosing energy from GreenPower programs does create business benefits and opportunities as well as helping the environment. Nearly 27,000 organisations are signed up for GreenPower.

If you sign up for a GreenPower electricity product you can:
- position your company as a good corporate citizen
- gain a leading 'green' edge over your competitors
- incorporate that product into a broader environmental policy
- advertise your use of GreenPower in your marketing communications and even use the GreenPower customer logo in some cases
- count your company's reduced greenhouse gas contributions as part of any legally required reductions

- improve your environmental performance rating, with the potential to be included in environmentally screened or socially responsible investment funds.

Research carried out by Sustainability Victoria found that nearly eight out of ten people would prefer to buy products from companies that contribute to GreenPower programs, rather than from those that don't. As a business, consider the marketing potential of purchasing GreenPower as well as the environmental benefits. Outside work, look out for other products from the increasing number of companies that use GreenPower. For more information about GreenPower go to www.greenpower.gov.au.

WORKER WELLBEING

Here are some simple tips for a healthy workplace and healthy planet:

- Use a lux meter to check lighting levels and make sure that they're both adequate and efficient.
- Make sure your HVAC systems are operated, cleaned and maintained properly, and that any filters are replaced as recommended. This is important for preventing moisture and mould problems, and diseases such as Legionnaire's disease.
- If you can, open windows to ventilate the office regularly. Don't rely on airconditioning intake for fresh air—according to the Green Building Council of Australia, most systems provide less than 10% fresh air.

> **Who is buying GreenPower?**
>
> - The Australian Museum
> - Dinosaur Designs studio and stores
> - Luna Media, publishers of *Cosmos* magazine
> - The Body Shop stores and office complexes
> - Fisher & Paykel
> - Westpac
> - Langridge Artist Colours
> - Insurance Australia Group Limited (IAG), which includes NRMA Insurance, SGIO, SGIC, CGU, Swann Insurance and ClearView Retirement Solutions
> - Corporate Express

- You may wish to consider an air cleaner or filter as part of your HVAC systems. See page 16 for information about how they work.
- Prevention is better than cure, so see if you can avoid bringing air pollutants into the workplace in the first place. Use only low-fume green cleaners, adhesives and interior paints. Look for those that are certified by the Good Environmental Choice program.
- New electronic office equipment, furniture and printed material can all 'off-gas' indoor air-polluting gases and fumes, such as volatile organic compounds. That 'new smell' of things like PVC chair mats is something you want to minimise. When you receive printed materials or new office furniture or other products that off-gas, make sure you regularly ventilate the rooms they are in for the first few months after getting them. Alternatively,

see if they can be taken out of any packaging and aired outside of the occupied office space, perhaps in a storage shed or garage, for a few weeks before being brought inside.

MAKING IT HAPPEN

In any workplace, it doesn't take much to change some practices if you make it easy for people. The bigger the workplace, the bigger the positive effect that small changes can have. If you have successfully made your home environment a greener place to live, talk to your boss about introducing simple but effective changes at work, but remember: there's not a lot of use in setting up an environmental program that no one uses.

Educating colleagues

Once you have new systems in place, you need to educate co-workers to make it easier for them to use the systems. Those who are less green-minded won't have the same commitment to changing their work habits as you do, so make it as easy for them as you can. Put up signs with energy-saving tips specific to particular equipment near the equipment itself. Clearly label any recycling bins and bins for reusable paper.

Hold an education session for staff and make it fun. If you put on a special lunch or afternoon tea, you're more likely to have everyone turn up. While they're eating, you'll have their undivided attention. Go through the systems that you have put in place and how staff should use them. Show examples of the signs you have put up and explain what they mean. Also outline the benefits for the environment and the importance of

Good Environmental Choice

Good Environmental Choice Australia is an independent eco-labelling program that helps consumers identify goods and services that meet certain high environmental standards. This eco-label assesses a wide range of potential impacts and considers the whole life cycle of the product. As well as the environmental impact of a product in its manufacture, use and disposal, the certification program also considers impacts on human health, including indoor air quality. For example, the standard for office furniture requires wood components to have low levels of formaldehyde and adhesives to be low in volatile organic compounds (VOCs). Office products that are Good Environmental Choice certified include:

- adhesives
- administration services and offices
- carpets
- cleaning products and services
- coatings and paints
- floor coverings
- furniture and fittings
- panel board
- paper
- plastic products, recycled
- printer and imaging equipment
- printers and printed matter
- printing inks
- publishers and published matter
- sanitary paper
- TV and audio visual.

greening up the business sector. With a bit of imagination you can turn it into an enjoyable team-building exercise.

Monitoring the changes

Document the 'before' and 'after' environmental assessment of your office. This will help you to measure the extent of waste reduction, changes in energy consumption and any savings made through greener purchasing or energy efficiency. Use this information to report on (hopefully) the success of the program to your boss, and suggest it be included in any promotional literature or corporate profile documents.

Tell your friends in other businesses, suppliers and contacts about your successes and encourage them to green up their offices. With any luck you might find that the greening of the business sector is contagious.

GREEN INVESTMENT

Put simply, investing ethically means integrating your personal values with your investment decisions. Those who yawn during the stock market reports on the television news may think that ethical investment considerations don't apply to them. Think again. If you have a superannuation fund, then effectively you are an investor. If you have any doubt about the investment power of superannuation, consider that by the end of 2010 total Australian superannuation assets had reached an estimated $1.32 trillion.

Recent changes to legislation mean that many Australian workers can now choose their super fund. Find out how your super fund invests. Through your fund you may unwittingly be investing in industries that pollute the planet or destroy ecosystems. If your fund doesn't consider environmental and other ethical issues, request that they change their product or switch to a greener fund.

Greener funds are no longer on the fringe of the investment market. VicSuper, one of Australia's largest public offer superannuation funds, with over 250,000 members, incorporates sustainability as a central operating principle. This is just one mainstream example. Choose to invest your hard-earned dollars in a way that makes you proud. Whether managing a large investment portfolio or simply deciding on a super fund, you can make choices that help to secure a better future for both your personal finances and the planet.

 For more information about ethical investment in Australia and New Zealand, visit the Responsible Investment Association Australasia website at www.responsibleinvestment.org.

MAKE A COMMITMENT

This week I will...

☐ _____

☐ _____

☐ _____

This month I will...

☐ _____

☐ _____

☐ _____

When I get the opportunity I will...

☐ _____

☐ _____

☐ _____

Notes

_____

_____

_____

_____

_____

_____

4

Green cleaning

In this chapter, we try to find a happy medium between the slobby, grotty approach favoured by some of the more relaxed housekeepers among us and the obsessively germ-phobic attitude of the neat freaks in our midst by looking at cleaning with the planet in mind.

Starting with the question of why we clean in the first place, we look at how we can combine the right hardware (cleaning tools, such as microfibre cloths) with the right software (simple green cleaning ingredients) to do many household cleaning jobs. Commercial cleaning products are then covered, and finally we look at pest control.

WHAT ARE WE AIMING FOR?

Cleaning is all about three things: health, appearance and maintenance. We do the cleaning we perceive to be necessary to make our home a healthy living environment. Aesthetics come into it, as we like our homes and clothing to look and smell their best. And in some cases, we clean objects and appliances so that they operate efficiently and last longer. But cleaning can involve a lot of unnecessary chemical cleaners and use a lot of water. We've all seen news stories about food poisoning, so we tend to get a bit obsessed with cleaning surfaces and killing germs. In fact, this can do more harm than good.

Watch TV advertising to see the many ways that advertisers portray bacteria and germs as miniature monsters, hell-bent on making cute babies sick. This is to make you think that you urgently need to buy the advertised products to protect your family from evil germs. The truth is that antibiotics and antibacterial cleaning agents, while vital where they are truly needed, are being grossly overused. In the healthcare system, antibiotics are being used so widely that some strains of bacteria are building up resistance to them. These 'superbugs' are effectively immune to common antibiotics. Prescribing antibiotics to fight infection is the 'nuke them' approach. It indiscriminately kills both the bacteria causing the infection and the beneficial bacteria that our bodies need for good health. The loss of 'good' bacteria is what leads to thrush, which some women develop after a course of antibiotics. In many cases, it is completely unnecessary. Antibiotics do not kill viruses, so they are completely useless in fighting flu viruses and rhinovirus, the causes of the common cold. Yet many people rush to the doctor and demand antibiotics for themselves or their children whenever they get a sore throat or sniffle.

The same happens in your home. Don't wage germ warfare, buying every antibacterial soap, toothpaste, spray, cleaner and other product under the sun. As they say, 'what doesn't destroy us makes us stronger'. Our bodies need the small challenges that a little bit of disease provides to build up strong, healthy immune systems. The 'hygiene hypothesis' theorises that the increases in asthma, allergies and some immune-system problems we're seeing in affluent societies are due to ultra-clean homes and a lack of early childhood exposure to microbes.

The key to a healthy home is removing or killing pathogens (the disease-causing germs) where they present a health risk, such as in food or on food preparation surfaces, and saying 'live and let live' to those bacteria that are beneficial (such as probiotics) or benign. Practise simple good hygiene, by washing your hands with soap and water after blowing your nose, after using the toilet and before meals. Use heat to kill the bacteria in food. Reduce the risk of food poisoning by cooking or reheating food thoroughly.

The common sense and science of cleaning

Cleaning aims to do several things. It removes unwanted organic matter (such as food spills), which breaks down and produces odours and sometimes toxic waste products that we want kept out of food.

> 'Soap, water and common sense are the best disinfectants.'
> William Osler, Canadian physician, 1849–1919

By removing this matter, we're also taking away potential food sources for pests, such as ants or rats. We're also aiming to control or remove conditions that otherwise might allow pathogens to multiply and thrive.

Bacteria, including pathogens, need warmth, moisture and food to survive and multiply. Washing hands with soap and water, and thoroughly drying them, physically removes both the germs and the moisture they need. Washing and drying dishes, clothes and household surfaces similarly remove much of the bacteria and their food sources. Bacteria are about 80–90% water and need moist conditions, so thorough drying is an important line of defence against pathogens. This is particularly important with kitchen sponges, which are often left damp and reused before they've dried out. Moisture is also the reason why salt and salty water can kill germs— the salt causes water to pass out of the bacterial cell. Bacteria are also pH sensitive, which is why some acidic liquids, like vinegar, have a mild sanitising effect.

Heat is also important in cleaning and hygiene. Heat improves the solubility of some substances, such as dried spilt coffee, or makes grease and fats more viscous and easier to remove. Bacteria are also highly temperature sensitive—they typically slow or stop multiplying when it gets too cold, which is why we keep some foods in the fridge, and start to die at temperatures of 60°C and higher.

Alcohol can also kill bacteria by dissolving their outer membrane or damaging their proteins. However, alcohol hand rubs strip natural oils from the skin and can cause irritations and rashes with too much use. Alcohol hand rubs are no replacement for washing hands with soap and water, but can be useful on an outing when there's no handy basin and soap.

Soap and detergent are useful because, in the simplest terms, oil and water don't mix. Some substances are 'polar' and dissolve in water; others, such as fats and oils, are 'non-polar' and don't dissolve in water. Soap and detergent molecules are very long and have a non-polar, oil-soluble end and a polar, water-soluble end. As such they're able to surround tiny droplets of oil, grease or wax and make the droplets soluble in water, enabling them to be washed away.

Enzymes may also enter the picture as biological cleaning agents. They break down larger, harder-to-dissolve substances (such as proteins in blood stains) into smaller, more soluble substances. Note that they're damaged by high temperatures, so hot washes will affect the performance of enzyme-containing laundry products.

Ultimately, cleaning science isn't a replacement for common sense. Prevention is better than cure, so we can do without harsher cleaning products by making sure we avoid making hard-to-remove messes in the first place. For example, wearing an apron while cooking can cut down on the effort needed in the laundry cleaning clothes. Regular light cleaning also makes more sense than putting off the job and ultimately making it more difficult.

GREEN CLEANING TOOL KIT: THE HARDWARE

As the previous section suggests, most household cleaning—including the use of most commercial cleaning products—involves combinations of tiny physical and chemical reactions to dissolve grease, remove dirt, neutralise odours and kill germs. However, some cleaning products have been developed that use few or no chemicals to do their job. The two main types are steam cleaners and fibre technology products.

Microfibre cleaning cloths are a recent advance in fibre technology. You can now choose from a range of polyester and polyamide microfibre cloth products designed with particular cleaning jobs in mind. Cloths have improved cleaning power because they use fibres of varying lengths, sizes and surface textures. Some are designed to pick up and hold dust particles; others are designed to be more abrasive, so that they can remove bathroom scum. With many of these products, you can control how thoroughly you clean by how wet you make the cloth. They don't need a lot of pressure. In fact, in my experience, they don't work as well when you really put the elbow grease in!

The best cloths and mitts generally don't require the use of detergents. Some particularly stubborn messes may need the help of a little detergent, but nowhere near as much as you would use with a conventional sponge. Chi Clean, Vileda and Enjo all produce ranges of cleaning cloths and mitts that use this advanced fibre technology. Mops with specially designed, padded microfibre mop heads are also available.

Good-quality microfibre cleaning cloths are a must-have for the greenie and non-greenie alike, as they're more effective and make cleaning easier than ordinary sponges, rags and cleaning cloths. In my opinion, the two essentials are a cloth for glass and a general-purpose cloth. There are also special cloths for car cleaning. Microfibre cleaning clothes should be washed periodically in hot soapy water, rinsed thoroughly and allowed to air dry. Don't wash them with fabric softener or bleach or dry them in a clothes dryer as this will damage the cloths or reduce their effectiveness.

Steam cleaners and steam mops are electrical appliances. You fill them with tap water, plug them into a power point and give them a few minutes to build up enough heat to produce steam through the cleaning head. As well as cleaning surfaces, the high temperature of the steam sanitises the surfaces. The most widely available types are steam mops, such as the Kenwood Steam Mop, and pressurised steam cleaners. Some are designed to also steam clean curtains, carpets, mattresses and upholstery. Small, hand-held steam cleaners are also available.

Other cleaning tools on supermarket shelves are disposable wipes and paper towels. Reusable cloths are preferable to disposable, provided they're thoroughly dried and not left damp to breed bacteria. Some spills can be cleaned up with old newspaper. For the odd job that needs paper towel, use a brand made with recycled content.

GREEN CLEANING TOOL KIT: THE SOFTWARE

There are six basic cleaning ingredients for general cleaning that you can buy from your local supermarket for very little money. Many of them form the basic components of much more expensive, heavily marketed cleaning products. Different combinations of these ingredients, along with a little water, some good-quality cleaning cloths, a regular cleaning habit and perhaps a couple of drops of aromatherapy oil, can replace most of the commercial cleaning products in your laundry or under-the-sink cupboard.

Seven uses for bicarb soda

1 Mix a little with water to make a paste to clean baked-on spills on your cooktop.
2 Place an open box in the back of your fridge to absorb food smells.
3 Remove stains from coffee and tea cups by rubbing them with a paste made from bicarb soda.
4 After cutting meat on a chopping board, scrub it clean with a brush and bicarb soda.
5 Sprinkle a little into the bottom of your rubbish bin to absorb odours.
6 After chopping onions, remove the onion smell from your hands by rubbing your hands with a bicarb soda paste while washing them.
7 Add some bicarb soda to some white vinegar in a bucket. Add hot water and mop the kitchen floors with it.

Simple green cleaning must-haves

Bicarbonate of soda (bicarb soda, sodium bicarbonate or baking soda) is abrasive and it softens hard water, so it cleans without scratching. It is also great at absorbing odours and is a mild disinfectant. There are dozens of uses for bicarb soda, both in cleaning and grooming, so buy it in bulk. For more information, see chapter 6, pages 85–6.

White vinegar Like all vinegar, white vinegar is slightly acidic, containing naturally occurring acetic acid. White vinegar dissolves grease, deodorises and is a mild disinfectant. It is great for cleaning glass and shiny surfaces, so keep some in a labelled spray bottle.

Lemon juice deodorises, is a mild bleach, cleans and is mildly acidic. Plus it grows on trees, literally.

Borax is a mineral found in nature and is usually shelved with other laundry products in the supermarket. Borax cleans, deodorises, is a mild bleach and a disinfectant, and helps to control some household pests. While borax is less harmful than many of the products it can replace, it is toxic and should be used sparingly. Borax is poisonous if eaten, so keep it out of reach of children and away from pets, and don't use it on food preparation surfaces. Borax can irritate the skin and eyes, so wear gloves if you have sensitive skin, and wear eye protection if you're spraying a solution that contains borax.

Pure soap is a 100% readily biodegradable general cleaner. It can be bought in bars or flakes, but dissolves better in flakes. You can make your own soap flakes by grating a bar of pure soap. Make sure you keep soap flakes dry, as they tend to clump.

Washing soda (sodium carbonate) cuts grease, disinfects, softens water and removes stains. Like bicarb soda, it also removes odours. It removes oil, grease and even wax without the fumes that normally go with many of the commercial products sold to do the same job. However, it can irritate the skin, so wear rubber gloves while you're using it.

Optional extras for special jobs
Tea-tree oil is a great natural cleaner and healer. Tea-tree oil disinfects and is anti-fungal.

Eucalyptus oil is very good at removing fat, grease and all manner of stickiness, as well as being great for relieving colds and the flu. However, eucalyptus oil is poisonous and can kill if swallowed, so keep it out of the reach of children.

Rosewater doesn't really do a lot, but it smells lovely.

Aromatherapy oils are relaxing or invigorating and smell a whole lot better than vinegar. Just pay attention to any precautions accompanying the oil and use as recommended.

GREEN CLEANING TIPS AND RECIPES
Floors with sealed surfaces
Mix ¼ cup of bicarb soda and ½ cup of vinegar in a bucket. Add warm water and a few drops of tea-tree oil. Use this solution to mop the floors. This is a great one to do with kids. They'll love the fizz created when the soda and the vinegar mix. A few drops of peppermint and grapefruit oils (or any favourite aromatherapy combination) can be added to freshen up the room and make the job more pleasant. Peppermint oil provides the added benefit of deterring ants.

An alternative is to leave the bicarb soda out of the mop solution and sprinkle some directly onto any stubborn spills or stains.

General-purpose cleaner
Mix 1 teaspoon of bicarb soda with 1 teaspoon of soap and a squeeze of lemon in ¼ litre of water. This is particularly good for cleaning bench tops. Use some extra bicarb soda on difficult spots.

General-purpose spray cleaner
Dissolve 2 teaspoons of borax and 1 heaped teaspoon of soap flakes in 3 cups of water and store in a labelled spray bottle.

General disinfectant
Mix ¼ cup of borax, ¼ cup of white vinegar and the juice of ½ a lemon in hot water.

Dishwasher powder
If your dishwasher is full of only lightly dirty dishes, put 2 teaspoons of bicarb soda in the powder tray. For sparkling clean glassware, use a dash of white vinegar as a rinse aid.

Removing smells
Body odour, cigarette smoke and other smells can be reduced or removed from clothing by adding a cup of bicarb soda to a washing load. You can also put a handful of bicarb soda into the bottom of your kitchen tidy bin or leave an open box in the fridge to control kitchen smells.

Laundry detergent
To make a cheap laundry detergent for lightly soiled loads, thoroughly dissolve ½ cup of pure soap flakes and ⅓ cup of washing soda in half a bucket of hot water. Add cool water to fill the bucket and mix well. It will set to form a soft gel. Use 2 cups per wash in a front-loader, 3 cups for a top-loader.

Carpet spills

Sponge off excess liquid, then saturate the remainder with soda water and sponge off again. Generously sprinkle the damp patch left with bicarb soda and allow it to dry. Vacuum the dry bicarb soda.

Grease and adhesives

Ever noticed the thick grease that collects on top of the range hood or above the stove? This can be removed with paper towel (with recycled content of course!) and eucalyptus oil. Eucalyptus oil is also good for removing the sticky patches left after you've taken off price tags and adhesive labels.

Showers

Soap scum can be cleaned off shower walls and doors using bicarb soda in the same way you would use a powder cleaner. Once the shower recess is clean, wipe the surfaces with a soft cloth with a little bit of baby oil or almond oil. The fine layer of oil makes it harder for soap scum to adhere to the wall and keeps the surfaces looking shiny. You can also add a favourite essential oil to the oil to fragrance it. Whenever you have a shower, the warm water will help to release the fragrance.

Windows

Clean windows with sprayed vinegar and a rag. If the window is particularly dirty, it may first need a clean with soapy water.

Toilet cleaner

Once a week give the toilet a rinse and quick brush with white vinegar.

Doona freshener

Freshen up doonas and blankets on a hot day. Put them on the line, lightly spray with rosewater, and dry in the sun. This will also air the bedding. Sunlight has a naturally sanitising effect, and the rosewater will give your bedding a delicate fragrance.

Mould

Apply diluted hydrogen peroxide to mould and leave for a few minutes. Wipe away with a damp cloth. White vinegar can also be used, but isn't quite so effective. Tea-tree oil is also good at removing mould, but it does have a strong smell and is more expensive. As with many cleaners and disinfectants, prevention is better than cure. Prevent mould from forming by regularly airing damp rooms and by using a fan when cooking or showering.

COMMERCIAL CLEANING PRODUCTS

In 2010, Australians spent well over $1.8 billion on household cleaning products, laundry products and pest control. This makes the cleaning product aisle at the supermarket a commercial battleground. In order to gain the upper hand in this battle for market share, each product promises to deliver whatever its manufacturer thinks its target customer desires, whether it be powerful cleaning performance, environmental responsibility, faster or easier cleaning, or just a cheaper price.

Walking down your local supermarket's cleaning aisle, just about every product seems to have an eco-, bio- or enviro-name, and makes a whole lot of scientific-sounding statements. For those of us without chemistry degrees, it can be hard to sort the facts from the hype and find out what works, what is good for the environment and what is necessary for your own personal needs. You may want to revisit the cleaning science covered earlier on pages 43–4, so that you can read between the lines in product advertising.

When people buy commercial cleaning products, they generally consider performance, price, environmental aspects and health aspects. These considerations can sometimes compete with each other. The following tips show what to look for and how particular contents affect the environment and/or your health.

Dirty money

At supermarkets in 2010, Australians spent:

- $157.3 million on air fresheners
- $60.4 million on disinfectants
- $181.4 million on dishwashing needs
- $214.3 million on household cleaners
- $810.2 million on laundry needs
- $195.3 million on pest-control products
- $262.2 million on mops, brushes, sponges, gloves and other cleaning tools

= a total of well over $1.8 billion spent on cleaning, laundry and pest-control products.

Source: *RetailWorld Annual Report 2010*

General cleaning products

Choose products that are certified by an independent environmental certification program, such as Good Environmental Choice Australia (www.geca.org.au). Brands with certified cleaning products include Orange Power, Herbon, Earth Renewable and Aware.

If you have asthma or allergies, consider Sensitive Choice–approved products. Sensitive Choice (www.sensitivechoice.com.au) is a labelling program run by the National Asthma Council Australia. The logo is awarded to products that have been reviewed by an advisory panel and found to offer a better choice to asthma and allergy sufferers. While not strictly an environmental labelling program, products that are suitable for allergy sufferers tend to have less environmental impact, so the program may be useful as a guide.

Choose products that are readily and inherently biodegradable, according to the Australian Standard (AS 4351). Favour those that state that all of the ingredients are biodegradable, not just the active ingredients. Better still, look for those that have met international standards of biodegradability (Readily Biodegradable Tests OECD 301A, 301E or International Standard ISO 7827).

Choose concentrated formulations in sensible packaging that can be recycled in your area. Avoid excessive packaging, such as individually wrapped portions.

If you have asthma or are sensitive to inhaled scents, avoid added fragrance. The word 'fragrance' on a product's ingredient list can cover dozens of different chemicals, many of which can irritate sensitive respiratory systems. Synthetic fragrances have been particularly implicated in inhalation allergies. If you like a little bit of fragrance, then look for products that are perfumed with plant-derived pure essential oils.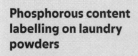

Avoid phosphates and phosphorous chemicals. Phosphates are added to laundry products and other cleaners and toiletries to soften hard water and make it more alkaline. However, phosphates cause algae to thrive, leading to algal blooms, which kill fish and other marine life and contaminate drinking water. Choose products that are completely phosphorous and phosphate free.

Avoid chlorine-based bleaches, containing chlorine or chlorine compounds (often hypochlorite salts). The manufacture of bleach often produces mercury and dioxins as waste or by-products, which are bad for the environment. Chlorine bleaches can also leave dioxin residues, which can accumulate in the body's tissues and are linked to a range of health problems.

> ### Phosphorous content labelling on laundry powders
>
> You may see a logo on laundry detergents that has either the letter 'P' or the letters 'NP' in a square box. These are the symbols of the ACCORD (originally the Australian Chemical Specialties Manufacturers Association) Phosphorus Standard, a voluntary scheme to regulate the phosphorus content and labelling of household laundry detergents. The 'NP' logo signifies 'no phosphorous' and indicates that the product contains 0.5% phosphorous or less. The 'P' logo signifies that the phosphorous content complies with the industry's agreed standard of 5% or less phosphorous concentration by weight in relation to the standard dose size. Look for products with the 'NP' logo or a statement that they are phosphate or phosphorous free.

Avoid petroleum-derived ingredients. Cleaning products and laundry washing liquids and powders often contain detergents or surfactants that are made from petroleum products. Petrochemicals are not renewable, are slow to biodegrade and can cause allergic reactions. Look for alternatives made with surfactants based on plant oils instead of those derived from petroleum oil (sometimes listed as mineral oil). These plant-based detergents break down quickly into harmless substances and are made from a renewable resource. Unfortunately, there is a wide range of petrochemical ingredients, making it difficult to identify them before buying a product. However, products that use plant oil–based surfactants will often state this clearly on the packaging because it's a selling point. Commonly used plant oils are coconut, palm kernel and corn oils. They may also be listed as vegetable oil.

Choose palm oil–containing products with care. While palm oil may be better for some sensitive skins than mineral oil, it does have a dark side. Rainforests in South-East Asia are being cleared to make way for palm oil plantations to keep up with the growing demand for palm oil, used in food and cleaning products and as a fuel. This is placing pressure on the remaining orang-utans and other wildlife that rely on these rainforests. Look for products that state that they are palm oil free or made with certified sustainable palm oil. For more information, see the Roundtable on Sustainable Palm Oil website at www.rspo.org.

Choose products with a cause. Some cleaning products help the planet by helping to raise funds for environment groups. This happens when a product manufacturer gives a green group some of the profits from sales. For example, the sales of Aware laundry products help to fund Planet Ark. Always make sure that the product is genuinely a better environmental alternative and that the donation to a green cause is not a just an exercise in greenwashing.

Favour pump-pack sprays over aerosols, particularly spray packs that offer refill packs. It takes huge amounts of energy to make and fill aerosol cans. In addition, because aerosols produce a finer mist that can more easily enter the lungs, they can pollute indoor air and cause respiratory irritation.

Avoid antibacterial cleaning agents. They kill both harmful and beneficial bacteria and are helping to create antibiotic-resistant 'superbugs'.

Trust your nose. Part of the reason we have a sense of smell is to alert us to possible danger. If a product has a strong, irritating odour, it may contain ingredients that are potentially harmful. If you don't like the smell, avoid it.

Choose products from companies that specialise in green cleaners. There are companies that cater specifically to the green market, such as Herbon, Earth Choice, Ecostore, Murchison-Hume and Orange Power.

Specific products

Air fresheners

- Don't buy air fresheners unless you absolutely have to. They are not necessary and have no cleaning effect. They simply give the impression of 'fresh air' to cover up poor indoor air quality. They can also mask other smells that indicate a real problem, such as a dead mouse or rotting food behind furniture, or bedding that needs airing.
- No product can replace regular ventilation to keep indoor air fresh. However, a few drops of peppermint, eucalyptus and/or citrus oils in your mop water can help to give your home a pleasing 'fresh' fragrance.

Dishwashing

- Avoid individually wrapped dishwasher detergent tablets, as this is unnecessary packaging, unless the wrapper is dissolvable and part of the formulation. Cardboard packaging made with recycled content is the environmentally preferred packaging for powder detergent. However, dishwasher detergent is caustic and highly dangerous when swallowed, so plastic bottles with childproof lids may be preferable if you have small children.

Furniture and floor polishes

- Avoid aerosol polish products, as aerosols are an energy-intensive and wasteful form of packaging. These polishes often contain a cocktail of solvents, petrochemicals, synthetic fragrances and other potentially harmful ingredients, such as nitrobenzene (classed as a likely human carcinogen). In addition, aerosols can be a very inefficient way to apply a product, as much of it is diffused into the air, where it contributes to poor indoor air quality.
- Look for furniture polish made from beeswax or plant waxes and oils, instead of petrochemicals.

Multipurpose cleaners

- If you are looking for a multipurpose cleaner that bleaches as well, choose one that is oxygen-based instead of chlorine-based.
- Choose orange or citrus cleaners over those based on petrochemicals. The active ingredient in citrus cleaners is limonene, a renewable, plant-derived solvent and degreaser extracted from the oil in orange peel. While much

less toxic and more environmentally friendly than caustic cleaners, citrus cleaners can still irritate people with sensitive respiratory systems. Limonene is also a volatile organic compound (VOC) and can pollute indoor air, so make sure you adequately ventilate your house during and after use.

- Some multipurpose cream cleaners boast the 'cleaning power of baking soda' (sodium bicarbonate, bicarb soda or bicarbonate of soda). Why spend the money on a commercial sodium bicarbonate cleaning product when plain bicarb soda is cheap to buy?

Oven cleaner

- Oven cleaners are generally made from either caustic soda (sodium hydroxide) or 'non-caustic' ethanolamine (also known as monoethanolamine). Caustic soda is highly corrosive and even in diluted form will severely irritate, and even burn, skin and eyes. Ethanolamine, while 'non-caustic', can cause headaches, a sore throat and respiratory problems, including asthma attacks. Oven cleaners may also be labelled as containing 'alkaline salts'. There are many different alkaline salts, with varying properties, so practically this statement has very little meaning. If you choose to use such products, always follow the safety instructions.

- Some oven cleaners offer a less caustic option, using citrus extract as the active ingredient. These oven cleaners are often more expensive, and a 2003 CHOICE test of oven cleaners found them ineffective.

> ### Read the instructions and don't overdose!
>
> When using commercial cleaning products, make sure you read the instructions—they are generally written to help you, not annoy you. Many cleaners give off toxic fumes or are unsafe to touch. If the label says to ventilate the room while cleaning, *do it!* If it says to wear gloves or protective clothing, *do it!*
>
> If the directions are unclear, ring the manufacturer to clarify. Better still, try to find a greener, less harmful alternative.
>
> Also, follow the instructions regarding dose size. For example, using twice the recommended amount of laundry detergent won't make your clothes twice as clean, but it will waste money and add an extra chemical load to the wastewater that is treated in your septic tank or your local sewage treatment plant.

- There is little to recommend in this category. However, the need for expensive and potentially harmful oven cleaners can be avoided. Simply give your oven a quick wipe with soapy water after each use. More persistent spills can be rubbed away with a damp cloth using a paste of bicarb soda and water.

Toilet products

- Don't use toilet cistern additives to colour and fragrance the water. They are essentially a cosmetic product. They add unnecessary chemicals to our waterways.

Laundry products

- Avoid optical brighteners and fluorescers. Optical whiteners are added to laundry products to give clothes that whiter-than-white brightness. They work by coating fabric fibres with white or slightly blue-white substances that reflect more white and so make the clothes look cleaner. They do not remove more dirt; they essentially whitewash it. Similarly, fluorescers work by absorbing ultraviolet light and re-emitting blue light to make the garment appear whiter. Producing and using unnecessary chemicals is not good for the environment, but another cause for concern with these products is that they can cause allergic reactions and skin rashes. People with sensitive skins should avoid them.

- Buy a concentrated product so you only need to use a small amount per wash. This means that you can wash more loads per box or bottle, reducing packaging waste. Also, fewer packages have to be made and transported to stores, so transport fuel and the associated greenhouse gases are reduced.

- Choose products formulated to work in cold water. That way you can capitalise on the energy savings of a cold-water wash, without poor results.

PEST CONTROL

'Control' is the operative word, not extermination!

We share the earth with a whole range of plants and creatures, and we are only just starting to understand the complex relationships between them and how they are interdependent. We have to keep the creepy-crawlies in our homes and gardens under control without harming the environment, other wildlife or our own health by using poisonous pesticides. This chapter looks at pests inside the home.

For more information on garden pest control, see chapter 11, pages 213–16.

Traditional pest control involved 'nuking' the pest with vast amounts of insecticides, bait and poison. While filming a movie on location in Australia, Hollywood actress Maureen O'Hara was filmed being sprayed with DDT to keep flies away. Little did we know then that the poisons that killed the flies and mosquitoes were also killing us.

Poisoning is a chemical means of pest control but there are alternatives.

The basics of green pest control

Enlist predatory allies. This method lets the animals that eat the pest do the job of keeping their numbers down for you. For example, leave some spiders that aren't harmful to humans (such as huntsmen and daddy-long-legs) around to catch other spiders, mosquitoes and other insects.

Be a home-wrecker. Most pests don't like to feel exposed. Limiting the number of sites where they can live or breed will help keep them out of your house. Ventilate your house in warmer months, as they tend not to like draughts.

Don't feed them. Pests go where there's food, so don't be surprised to get a pest problem if you're in the habit of leaving food out or not cleaning properly. Clean up food spills quickly and avoid eating in bedrooms, particularly if they are carpeted. Keep garbage in bins with well-fitting lids and wash them regularly.

Indoor and outdoor products

Never use a garden pesticide or insecticide indoors unless the product label states that it is safe to do so. Chemicals for use in the yard assume that any harmful gases will be quickly diffused in outside air. However, harmful gases stay in harmful concentrations indoors, particularly in homes with poor ventilation.

Put up barriers. Physical barriers will help to keep pests out. Block cracks, holes and gaps that allow entry into your house. These gaps are often found around windows and plumbing and electrical outlets. Termite mesh is also an option for smaller creepy-crawlies.

Get some chemical help. Some substances are non-toxic to humans but poisonous to pests. Others simply repel them. Strategically place repellents around the house to keep pests away or, if all else fails, use a non-toxic insecticide.

Use brute force. Another option is to kill pests using old-fashioned traps and a trusty fly swat. You can also get sticky traps to catch small insects, such as pantry moths.

Be aware of seasonal variation. The pest problem may decline or be cyclic, particularly in the cases of mice, spiders, millipedes and ants.

Pest control tips—critter by critter

Ants Don't leave out uncovered food or food spills, and wipe out your oven or griller after use. Keep your garbage bins away from the house if possible. Ants also like pet food, so be careful where you place your pet's bowl.

Mix something sweet, such as jam or honey, with borax to make an effective ant poison. Put the mixture on small saucers or container lids and place them near ant trails or in cupboards where you have an ant problem. Make sure children know not to eat the mixture, and don't use it if you have small children or pets that may unknowingly eat it. If you find the ants' nest, you can kill the ants and destroy their nest by pouring boiling water into it, although consider this a last resort as ants are important for soil health.

Ant-repelling herbs such as pennyroyal, rue or tansy planted in small, strategically placed pots or dried bunches provide a less gruesome way to keep ants away. You can also try to deter them by sprinkling a little cayenne pepper at the points where they enter the house. Mop floors with soapy water with a few drops of peppermint oil to erase the scent trail and deter the ants.

Cockroaches Keep cockroaches out of the house in the first place by sealing cracks and crevices in walls and openings around pipes. Make sure your doors and windows fit snugly. Cockroaches are repelled by cucumber (put thin slices or the skins on paper towel in the back of the cupboard and replace every week or two), bay leaves, pyrethrum and vanilla beans. Try leaving these in cupboards that regularly get invaded, or wipe them with a little tea-tree, peppermint or citronella oil.

Bait cockroaches with sugar laced with 5% borax, but keep this out of the way of children and pets. You can also make a trap from a glass jar that has been greased on the inside and half-filled with beer. The cockroaches won't be able to resist the offer of free beer and will fall into the jar, be unable to escape and drown. At least they'll die happy. If you must use an insecticide, use one based on natural pyrethrum.

Dust mites can cause asthma attacks, but their numbers can be reduced by regularly washing bedding, curtains and upholstery, and vacuuming carpets. You may wish to remove carpeting from your house altogether, or at least from the bedrooms. Dust mites are sensitive to extremes in temperature and sunlight. Wash bedding at least once a fortnight in a hot water wash of at least 55°C, adding a little eucalyptus oil to the wash cycle and drying outside in the sunlight. Air blankets and doonas in sunlight.

You can make your indoor air quality healthier and you can reduce the likelihood of anyone having an asthma attack in your

Actions to decrease dust mite numbers

- Keep bedrooms well ventilated and well lit with natural light.
- Use mattress, doona and pillow protectors, and regularly clean the covers.
- Wash sheets, pillow cases and quilts in hot washes (at least 55°C or 131°F).
- Air blankets and doonas weekly, preferably in sunlight.
- Regularly vacuum mattresses, furniture and curtains as well as carpets.
- Invest in a vacuum cleaner with strong suction and a good filter system.
- Dust with a damp cloth instead of a feather duster.
- Limit the number of open storage areas in the bedroom where dust can collect, such as under the bed, on top of wardrobes and open bookcases.

household by taking measures to reduce dust mite numbers in your home. This is particularly important in the bedrooms of any small children. Dust mites are sensitive to dry conditions, changes in temperature and sunlight (particularly ultraviolet light). These can be used to control them.

Flies Again, as with most pests, don't leave out food or food scraps, as they attract the flies. Fit flywire screens to your windows and doors to keep them and other insects out of the house. Trap flies using traditional sticky flypaper (available from supermarkets and hardware stores). Sharpen your aim and reflexes by killing flies with a fly swat. Grow insect-repelling plants such as eau-de-cologne mint or pyrethrum near doors and windows.

Mosquitoes Most people think of mosquitoes and their bites as just an annoyance. However, mosquitoes can also spread viruses, such as Ross River fever.

Mosquitoes become a particular problem when you've given them some nice still water to breed in, so get rid of stagnant water around the house. Make sure that any buckets or containers left outside are under cover from rain or left upside down, empty your pet's water bowl at the end of each day and refill it instead of just topping it up, change the water in wading pools or birdbaths twice weekly, and make sure water features or ponds in the yard have either moving water or a stock of fish.

When you're entertaining outdoors, burn citronella candles to repel mosquitoes. Alternatively, make a personal repellent of one part lavender oil, one part eucalyptus oil and one part pennyroyal oil with three parts of a carrier such as moisturiser or almond oil. Combine well and rub onto exposed skin.

Moths There are a number of natural fragrances that humans enjoy but moths will avoid. Make muslin sachets of dried lavender, orange or other citrus peel, rosemary, cloves or mint (or a combination of these), and place them in your linen cupboard and drawers and hang them in your wardrobe. Lavender and rosemary are both relaxing scents, so place muslin bags of these in with your bed linen. Eucalyptus oil and Epsom salts also repel moths.

The natural camphor in camphor wood is also great for keeping away moths and other insects, which is partly why it has been used in the past to make hope chests and glory boxes. Small balls made from camphor wood are also available to use in place of mothballs. For more information on why not to use mothballs, see chapter 5, page 64.

Pantry moths can be caught using fairly inexpensive pantry moth traps, such as those from EnvironSafe Products. These small, tent-shaped sticky traps have a natural attractant and are pesticide-free.

Rodents Keep benches and cupboards free from spilt or exposed food. Make sure food such as cereals, flour and chocolate (not surprisingly, a favourite with mice) is kept in sealed containers. Check your walls and floors for gaps through which they may be entering. Seal the gaps: steel wool is a good, unfriendly interim measure. Use old-fashioned mousetraps instead of poison to kill mice. Poison is more harmful to the environment and is toxic for children and pets. Bait the mousetrap with a small bit of chocolate or peanut butter rather than cheese. It's much more effective.

Silverfish like warm, damp conditions such as the areas around pipes that carry warm water. Regularly vacuum such areas to remove food particles and silverfish eggs. You can also catch them in homemade traps. Simply wrap clear or masking tape around the outside of a glass jar to provide a surface that silverfish can grip to. Put some cereal inside the jar to act as bait. They will climb up the sides and fall into the jar but will be unable to climb the smooth inside of the jar to escape. Trapped silverfish can either be let out outside, away from the house, or drowned in soapy water.

Spiders Get to know spiders to find out which ones you want and which ones you don't. Leave the daddy-long-legs. Their pincers can't pierce human skin, but they are highly venomous. When they fight with other spiders, they win, so a couple of daddy-long-legs will help you to keep the nastier spiders away. You can limit their numbers by making sure you regularly clear the house of cobwebs. The Australian Museum has a section on their website devoted to Australia's spiders. Visit australianmuseum.net.au/spiders to help you identify which of these eight-legged critters are dangerous.

MAKE A COMMITMENT

This week I will...

☐ _____

☐ _____

☐ _____

This month I will...

☐ _____

☐ _____

☐ _____

When I get the opportunity I will...

☐ _____

☐ _____

☐ _____

Notes

_____

_____

_____

_____

_____

_____

5

Green wardrobe and linen press

Everybody needs clothes. Clothes keep us warm, sheltered and modest. Bedding, towels and soft furnishings also serve a role in making our homes more comfortable. But fashion, be it clothing or interiors, has typically shied away from sustainability considerations. The textile industry is fraught with challenges. Growing fibres and dyeing them, or producing them from petrochemicals, consumes and pollutes energy, water and resources. More broadly, the fashion industry thrives on the fast turnover of strong trends that start to look dated long before the clothes or furnishings have worn out—in fundamental opposition to the principles of green consumerism. Then, beyond the environment, there are the social considerations of offshore manufacturing, sweatshop labour and child labour. It's little wonder that the industry tends to put ethics in the 'too hard' basket.

Granted, a handful of clothing companies have ventured towards lower-impact clothing, often manufactured with natural fibres, based on the assumption that natural fibres and products are always environmentally preferable. Such products have tended to look more hippie than hip, catering to a niche market rather than the mainstream.

Slowly but surely, the textile, clothing and footwear industry is starting to face its ethical and environmental challenges. A new breed of fashion designer is emerging: one who considers environmental impacts and tries to reduce them while still maintaining a distinct design aesthetic. This field is still very new and is more developed overseas. So while the choice in Australia's retail stores is currently limited, it will expand in coming years. This chapter looks at the eco-options for dressing your body and stocking your linen press.

TEXTILES AND THE ENVIRONMENT

Many people believe that the greenest fabrics are natural fibres. As with cosmetics, 'natural' doesn't necessarily mean 'environmentally friendly'. Even fabrics made with natural fibres can have vastly varying environmental impacts. Some require more water and pesticides than others; some produce polluted wastewater during processing; some are made from non-renewable resources; some are more easily dyed, and some are easily laundered while others require dry-cleaning. Toxic chemicals are used in cotton farming and in the processing and manufacturing of many fabric types. Many synthetic fabrics, however, are not biodegradable. Fabric production is also a big water user and potentially a contributor to water pollution.

Clothes are also a growing contribution to landfill, particularly with the growth of 'fast fashion' clothing with a short useful life span. In 2000 alone, the world's consumers spent around US$1 trillion worldwide on clothes. Retail brands such as Sportsgirl and Zara reinvent their image every six months (each fashion 'season') to stimulate sales.

The fast fashion phenomenon is also evident in interiors and soft furnishings. Interior trends used to change in slower timeframes than clothing fashions. Now furniture retailers such as Freedom Furniture are also launching a different look each season, in the same way the apparel industry launches a new collection every six months. Each newly launched collection automatically devalues the previous season's products in the eyes of the fashion-conscious consumer, stimulating the desire to replace garments and furnishings that are still in good condition. But you don't have to stay on the fashion treadmill, slavishly buying each new trend. You can express your individuality, and put together a good-looking outfit and an attractive home, without it costing the earth! Yves Saint Laurent himself put it best: 'Fashions fade; style is eternal'.

CLOTHING

Fortunately, the fashion industry is starting to improve. This is partly because its primary suppliers, wool farmers for example, are looking at how they can improve their own practices. They're also seeing that they can

> It takes around a third of a kilogram of synthetic fertilisers and pesticides to produce the average pair of jeans from conventional cotton.

gain an edge over their competitors by offering clothes made from greener fabrics. The good news is that the eco-chic trend is growing right across the clothing industry, from mainstream casual fashion to active sportswear to designer labels. For example, both Levi-Strauss and Nike have increased their use of organic cotton and are blending it with conventional cotton. Giorgio Armani himself, fashion's emperor of up-market good taste, is using hemp in some of his clothing ranges. Edun is a socially conscious clothing label run by Ali Hewson, her famous husband Bono and New York fashion designer Rogan Gregory. Though the emphasis of Edun is on fair trade and sustainable employment in the developing world, around 30% of the cotton it uses is organic. These labels are among the style leaders that mainstream fashion follows, so this could be the beginning of the greening of fashion.

Closer to home, Gorman, Bird Textiles, Sosume and a handful of other fashion brands are pioneering eco-fashion in Australia. But there's much more to a greener wardrobe than going out and buying new clothes, no matter how eco-friendly they claim to be. Reducing the ecological footprint of your wardrobe is about how you care for—and make best use of—the clothes you have now and those you buy.

Choosing and caring for your wardrobe

Less is more. Why do we always find ourselves with a wardrobe full of clothes and nothing to wear? These wardrobes are often full of cheap, fashionable impulse purchases that were red-hot the day they landed in the store but are now out of fashion or were so poorly made that they haven't stayed in good condition. Any wardrobe consultant, environmentally aware or not, will tell you that it's better to buy a few well-made, basic garments with your hard-earned dollar than an armload of statement-making fashion pieces.

Go for alternative fibres. Look for garments made from organic cotton or hemp fibre and help to create a market for farmers who grow these greener crops.

Buy garments with easy care instructions. Dry-cleaning leaves residual solvents in the garment that are emitted into the air at home, contributing to unhealthy, poor indoor air quality. Wherever possible, avoid buying 'dry-clean only' garments.

Share and share alike. Swap and share clothes with friends who have similar tastes and are a similar size. This swap system only falls down when someone fails to look after the clothes they borrow, returns them with obvious stains or forgets to return borrowed items promptly.

> **From soft-drink bottles to fabric**
>
> In the USA, recycled PET is used to make fibrefill—a padding fabric. Five recycled 2-litre soft-drink bottles make enough fibre fill for a man's ski jacket and thirty-five 2-litre bottles make enough fibrefill to fill a sleeping bag.

Buy recycled. Some Polarfleece clothes from the USA are made from plastic recycled from soft-drink bottle plastic. Look out for garments made from fabrics marketed as 'Synchilla', 'Ecospun' or 'Ecofleece', such as Kathmandu's 'Stromboli' jackets (made from Ecofleece). Occasionally, local designers will also include recycled plastic fabrics, such as the Billabong board shorts made from 'eco-recycled surf satin'.

$ Go vintage and second-hand. You can get some great retro clothing for a fraction of the price from charity stores and recycled clothing boutiques. Buying second-hand clothes and other products makes better use of our resources and reduces the amount of waste being sent to landfill. When the second-hand store is run by a charity, the money you spend there goes to a worthy cause. You never know: you may even unearth a valuable vintage designer original. Second-hand trading sites, like eBay, can also be a great resource for finding pre-loved fashion.

$ Look after your clothes. You can keep your clothes looking newer for longer just by looking after them. Wear old clothes when you're doing housework or other grubby activities, do the fifties housewife thing and wear an apron to protect your clothes while cooking, use a front-loading washing machine (because it's gentler on clothes than a top-loader) and mend minor tears or holes.

Repair and alter. Try to do any minor repair work yourself, such as re-sewing buttons or re-sewing a hem that may have come down in places. Alternatively, some dry-cleaning businesses also do repair and alteration work at a fraction of the price of a new garment, particularly if the garment in question is good quality and therefore more expensive to replace. This will help to give a garment a longer life. If you find yourself needing a new wardrobe because you've lost or gained weight or your figure has otherwise changed significantly, see if existing clothes can be professionally altered.

Go for green brands. Favour labels that have good environmental policies. For example, accessories and homewares company Dinosaur Designs are carbon neutral, buy green electricity, recycle within their studio, buy recycled-content paper and have had water and energy audits conducted on their studio to improve their performance.

Moth-proof your wardrobe

There's nothing worse than pulling on a garment while you're in a hurry to get dressed and finding that it's been moth-eaten. In the past, the unpleasant alternative was to fill your wardrobe with smelly mothballs. Luckily, mothballs have gone out of fashion as they are not a good green alternative for pest control. They are a petroleum-derived product made with a cocktail of synthetic chemicals, including naphthalene and para-dichlorobenzene, both classed as possible human carcinogens. They also bear an unfortunate resemblance to Kool Mints and so should never be left where a child might find them and eat them.

There are natural alternatives to mothballs. Camphor wood contains natural oils that repel insects. You can buy camphor wood balls to replace mothballs. You can put jumpers and spare bedding into old-fashioned 'hope chests' made from camphor wood. You can also fill small muslin bags with dried herbs that repel moths. Use a combination of dried lavender, orange peel, rosemary and mint and hang them in your wardrobe. You may like to keep a spare sachet of lavender and rosemary near your bed, as these scents are relaxing fragrances.

Consider hiring. Consider hiring garments that you expect to only wear once or twice, such as bridesmaid dresses and tuxedos for weddings. There are also a small number of companies that rent out designer handbags and accessories, such as Love Me and Leave Me (www.lovemeandleaveme.com) and Bag an Image (www.baganimage.com.au).

Hold a swap party. Get together with friends and hold a clothes-swapping party. Each person brings the clothes they want to give away. Then guests take it in turns to choose items to keep from the pooled garments, with the remainder donated to charity. Swap parties are particularly useful among families with children who often out-grow a piece of clothing while it is still in good condition.

Go to a clothes-swapping event. Taking the swap party to a whole new level, some of Australia's capital cities have the occasional major clothes-swapping event, attracting eco-fashionistas from near and far. Organisations that run events like these include the Clothing Exchange (clothingexchange. com.au) and Swap 'Til You Drop (www.swaptilyoudrop.com.au).

Swap clothes online. The trend has gone online with Thread Swap, a website where you can gain credits for clothes you offer and use the credits to 'buy' clothes from other swappers. A small fee is charged for exchanges. See www.threadswap.com.au for details.

→ **ASK TANYA**

I saw an eel-skin purse in a boutique and almost instantly I was wary of its sustainable credentials. I found an article stating that the eel skin trade in Korea is quite sustainable as eels are eaten in Korea so the skin is a by-product. Please enlighten me!

Deanne, Wantirna South, Victoria

This reminds me of the 'should vegetarians wear leather shoes' debate. In each case, the skin is a slaughterhouse by-product, which might otherwise be sent to landfill. Some strict vegans take the view that humans shouldn't use any animal products at all. Others take the view that if an animal life is taken (and it's unlikely the world will turn vegan in the immediate future), we have a moral obligation to make as much use of it as possible. There's a food movement called 'nose to tail eating' based on this thinking.

Just remember that there were other environmental impacts stemming from turning the skin into a purse and ask yourself if you have a genuine need for a new purse in the first place. With all eco-fashion products there's a temptation to justify buying something 'because it's green', but the real environmental benefit comes when the purchase of the eco-fashion item displaces the purchase of a more detrimental alternative. The planet is rarely helped by buying more stuff on top of what we're already consuming. To me, the bigger question is whether or not eel meat is sustainable as stocks are collapsing around the world. But while it is still being consumed, the potentially useful skin by-product remains available.

Eco fashion: what's available

Buying at least some clothing or manchester made from sustainable fibres will help the development of these alternatives, hopefully displacing some of the more harmful fabrics in the long term. In Australia, eco-fashion choices are currently limited. Here is a selection of what's available:

- Designer Lisa **Gorman** has been a leader in high-end fashion, applying sustainability principles to her label Gorman. She also has Gorman Organic—a range of basics made from organically grown fibres. Gorman and Gorman Organic garments are available from Gorman stores or boutique stockists. See www.gorman.ws for details.
- Beautiful eco-garments and fabrics from **Bird Textile**, with fun retro prints and styling, can be bought from a number of stockists around the country. Stockists are listed on their website at www.birdtextile.com.
- **Hunter Gatherer** is a retail fashion venture of the Brotherhood of St Laurence, a non-profit organisation. It's really an op shop with a difference. Two Hunter Gatherer stores in Melbourne (Fitzroy and Melbourne City) sell the coolest vintage and retro clothes sourced through the Brotherhood's charitable collections. The range of second-hand clothes is supplemented with their own range of new 'vintage inspired' garments, made locally and accredited with the No Sweat Shop label. For store locations, see www.bsl.org.au/hunter-gatherer.aspx.
- Australian company **Skin and Threads** (www.skinandthreads.com) uses organic cotton for their womenswear range, which they describe as 'designer basics'. Their website lists dozens of stockists across Australia and the USA.
- **Skinny Nelson** (www.skinnynelson.com.au) has organic cotton men's and women's clothes. The range is young, edgy and funky, with mostly black, white and grey T-shirts, leggings, jumpers and tops.
- Another up-market, sophisticated womenswear label is **Sosume** (www.sosumeclothing.com).
- **Sara Victoria** (www.saravictoria.com.au) has an 'Organic Softwears' collection of designer womenswear, made from fibres including organic cotton, hemp, wool, silk and flax.
- 'Slow fashion' label **Pure Pod** (www.purepod.com.au) has womenswear separates with designs inspired by nature.
- **Maud N Lil** (www.maudnlil.com.au) produce men's and women's ranges of organic cotton underwear, as well as organic cotton soft toys.
- **Gaia** is another Australian organic cotton apparel company, offering baby clothes. Visit www.gaiaorganiccotton.com.au.

Dry-cleaning

Dry-cleaning uses solvents instead of water to clean fabrics. This is partic-ularly useful with fibres that shrink, roughen or deteriorate when washed in water. Conventional dry-cleaning uses the solvent tetrachloroethylene, also known as perchloroethylene, or 'perc', which is listed by WorkSafe Australia as a 'suspected carcinogen'. Dry-cleaning workers are exposed to perc on a daily basis. When dry-cleaned clothes are brought home, they con-tinue to 'off-gas' solvent residue, exposing the whole household to potentially harmful gases.

If you have clothes that must be dry-cleaned, ask your dry-cleaner about their methods and solvents. There are greener alternatives to perc. The three most common alternatives are as follows:

- **Wet cleaning** uses water and biodegradable detergents in very gentle specialised equipment. The humidity is also generally controlled during the drying process, and pressing and tensioning equipment is used to prevent shrinkage and to maintain garment shape. Wet cleaning is the method used by the dry-cleaning chain Daisy (www.daisy.net.au).
- **Carbon dioxide cleaning** is another non-toxic alternative to perc dry-cleaning. It turns out that when carbon dioxide is put under enough pressure to make it a liquid, it behaves like a solvent. The liquid carbon dioxide is passed through the fabrics and removed. Once the pressure is released, it returns to a gaseous state and any dirt and other waste simply drops out. The carbon dioxide used is a waste product of alcohol produc-tion, temporarily diverted for use in dry-cleaning, so there is no increased greenhouse contribution.
- **Siloxane** is liquid silicone used as a solvent in dry-cleaning, including in a product marketed as 'GreenEarth'. It's made from sand (which, while non-renewable, is plentiful), is free from volatile organic compounds (VOCs), and is thought to be non-toxic in the levels found in dry-cleaning shops and as residues in clothes. However, there are some question marks over its life-cycle environmental impacts. There are concerns about the possible formation of formaldehyde in the breakdown prod-ucts, and its manufacture uses chlorine. A Greenpeace USA study into toxic substances in dry-cleaning noted, 'Given that these processes also involve heat, oxygen, and often copper catalysts, it is likely that dioxin and other organochlorine compounds are released during production either as emissions or from burning in production waste incinerators. Thus, the life cycle of siloxanes could have, at root, the same problem as perc'.

MANCHESTER: GREENER OPTIONS

Organic cotton bed linen and towels are an alternative that is less harmful to the environment. Global concern for soil and land use, along with unease about pesticides and chemical fertilisers, has led to a boom in organic farming. Cotton crops are thirsty and prone to pests, so they need large amounts of water and pesticides. Given that Australia is a dry continent, many people think that Australian cotton farms should instead be growing industrial hemp. Australian brands of organic cotton manchester include ecoLinen (www.ecolinen.com.au), Organature (www.organature.com.au), Blessed Earth (www.blessedearth.com.au) and EcoDownUnder (www.ecodownunder.com).

Conventional cotton doesn't take dye well. If you're considering buying cotton products, then, as a rough guide, darker coloured bed linen will have needed more chemicals to dye the fabric and fix the dye. The dyeing process also produces polluted wastewater. On the plus side, however, cotton is fairly easy to care for and generally doesn't need dry-cleaning. As well as making organic cotton bedding, many alternative bedding companies also offer undyed conventional cotton bedding.

Hemp is a strong, durable fibre and a fast-growing crop. Hemp crops have few pests and need less water and fewer pesticides than conventional cotton crops. One of the other benefits of hemp fibre is that it holds dye very well, unlike cotton, so hemp fabric can be dyed in a wide range of vibrant colours, using less dye and associated chemicals in the dyeing process. Hemp has had a little bit of bad (or in some cases, good) PR, because of its association with marijuana use. Agricultural hemp, however, has negligible amounts of the narcotic tetrahydrocannabinol (THC), which is what gives marijuana its more recreational properties. Hemp crops can be used to make a range of products, from bedding to hemp oil to cosmetics to rope and string. Hemp bedding brands include Margaret River Hemp Co. (www.hempco.net.au), Hemp Gallery (www.hempgallery.com.au) and Made in Hemp (www.madeinhemp.com.au). Eco-retailers, such as the online store Biome (www.biome.com.au), also stock hemp and organic cotton blend bed linen products.

Wool can help you to reduce your greenhouse emissions, though sheep farming and wool scouring does have an impact on the environment. Wool is very effective at retaining heat. Using wool blankets on your bed will take away the need to use additional heating in your bedroom. This will save energy and the greenhouse gas emissions associated with using energy.

With worldwide interest in environmental issues, organic and carbon-neutral wool products are now becoming available.

Bamboo fibre is essentially a cellulose fibre, like rayon or viscose. It can be made by either of two methods. The chemical method extracts cellulose from bamboo pulp using one chemical and regenerates it using another, but it does use hazardous carbon disulphide. Alternatively, the mechanical method crushes stems and uses enzymes to separate the fibres. As a material resource, bamboo has some significant sustainability advantages. Harvesting involves cutting lengths of bamboo stems, without disturbing the soil and root system. Like other grasses, it regenerates quickly and doesn't need replanting, unlike hemp or cotton, or traditional timber forestry. Many bedding stores and eco-retailers now offer bamboo-fibre bed linen, towels and bathrobes. In coming years, look for bamboo fibre made without carbon disulphide.

> ### Warmer with wool—above and below
>
> Many people wonder why they're still cold, even with extra blankets and quilts piled high on top of them. Extra bedding on top stops heat loss from that side of your body, but it doesn't stop the heat lost from underneath your body. Don't forget to insulate underneath as well as on top by using a wool underlay, extra blankets and/or a thicker mattress protector underneath the sheet covering your mattress.

Synthetics are sourced from petroleum, of which there is a limited supply, so synthetic fibres are not a renewable product in the long term. Like all petrochemical products, they do take a toll on the environment and contribute to the greenhouse effect. They are also more flammable than natural fibres, and tend to be less absorbent and to emit small amounts of air-polluting fumes. However, they can have less environmental impact from washing and ironing.

MAKE A COMMITMENT

This week I will...

☐ _____

☐ _____

☐ _____

This month I will...

☐ _____

☐ _____

☐ _____

When I get the opportunity I will...

☐ _____

☐ _____

☐ _____

Notes

_____

_____

_____

_____

_____

6

Green grooming and bathing

Humans generally care about how they look and smell. Our skin and hair are central to our appearance. Everyone wants to be as beautiful or handsome as possible. Indeed, our use of cosmetics dates back to the ancient Egyptian, Greek and Roman civilisations. Now in the twenty-first century, Australians annually spend in excess of $6.3 billion on cosmetics, perfumes, personal care products and toiletries.

Aesthetics aside, your skin and hair perform important functions. The skin is the largest organ of the human body, accounting for around 15% of a person's weight. It's your body's interface with its surroundings, protecting your more delicate tissues from pathogens, providing insulation and regulating your temperature. Skin also performs the tasks of gathering information through the sense of touch and of manufacturing vitamin D in the presence of sunlight (important for healthy bones). Hair also plays a role in regulating body temperature.

Some personal care products perform a function in caring for our health. Washing your hands with soap and water is fundamental to good hygiene. Brushing your teeth is part of good dental care. Skincare can help to control bad cases of acne, reducing infection and the load on the immune system. Then there are cosmetics. Cosmetics, such as lipstick or concealer, are personal care products that aim to enhance a person's appearance. Other toiletries, such as bubble bath, are simply luxury items used to 'pamper' ourselves. They feel good to use, and there's no crime in wanting that. But as much as we all want to look good and feel good, it is very easy to cross the line into gross overconsumption in the pursuit of youth, beauty and luxury.

Millions of dollars are spent marketing youth, beauty and luxury to us, primarily because the global cosmetics and fragrance business has an estimated annual turnover of US$170 billion. Our challenge as green consumers is to find a balance between looking after our health and enjoying a little indulgence, while resisting the temptation to overindulge.

This chapter will look at personal care, understanding product advertising, buying grooming products and making your own skincare preparations. It will also discuss sunscreens and animal welfare issues.

GREEN IN THE BATHROOM

Being green in the bathroom isn't only about vanity. Our day-to-day bathing and hygiene habits can have an impact on the environment, aside from the water use (as discussed in chapter 10). Here are some general tips for greening up the bathroom:

- Avoid over-packaged toiletries and bathroom products. Bars of soap do not need to be individually wrapped. Toothpaste is toothpaste whether it comes out of a simple tube or a pump that has used up more resources in its production.
- Use a cloth face washer rather than a natural or synthetic sponge. The harvesting of natural sponges disturbs the marine environment, and synthetic sponges are generally made from non-sustainable plastics.
- Choose reusable instead of disposable products. For example, don't buy plastic disposable razors; instead, get metal razors with replaceable blades. The Colgate and Persona brands both offer a toothbrush with a replaceable head.
- Reduce the need to use chemical drain cleaners by keeping the plughole in the shower or bath clear of soap scraps and hair. If possible, sit a sink strainer over the plughole.
- If weather permits, throw open a window in the toilet when you need to, instead of using an air freshener.
- Use a fan when using the bath or shower to remove moisture and prevent the growth of mould.
- Use the rubbish bin, not the toilet, to dispose of small bits of rubbish such as cotton tips and tampon wrappers.

→ **ASK TANYA**

I have a question, which I don't recall ever being discussed anywhere—tampons versus pads. An embarrassing issue for some but hey, it is a fact of life. So, what is best? Pads usually contain more product but some have less wrapping. Tampons, well, ones with applicators or ones without? Has there been any real research done on this, and if so, what were the findings? Also, do any of the brands on the market in Australia fulfil obligations such as using Fairtrade components and/ or organic?

Michelle, Worongary, Queensland

Sweden has given us more than ABBA and IKEA: researchers at the Royal Institute of Technology in Stockholm published the Comparative Life Cycle Assessment of Sanitary Pads and Tampons *findings in 2006. Their assessment covered human health, ecosystem quality and resource use impacts including land use, eco-toxicity, use of fossil fuels and contribution to climate change. The authors noted that there were data gaps but put tampons ahead of pads, with the disclaimer that more information on the inputs of cotton transportation and the energy use of tampon production might change the result. So where does that leave us? On the face of it, tampons are greener than pads on average, but there are some alternative feminine hygiene products that are better than 'average'.*

Reduce material use and waste by not getting tampons with applicators, or pads that are individually wrapped (a clean toiletries bag can keep them neat and clean in a handbag, briefcase or luggage). Look for unbleached products to avoid the impacts of chlorine bleach and the risk of residue dioxin in the pad or tampon. Certified organic cotton tampons and pads are available from health stores and eco-retailers. Remember that to reduce the risk of toxic shock syndrome, tampons can't be used overnight. A local Tasmanian company Eenee (famous for their nappies) makes biodegradable and flushable pads. Washable, and therefore reusable, cloth pads are also on the market (one has the dodgy-sounding name of 'Pleasure Puss'). Menstrual cups are also an option, though hard to come by in Australia. A menstrual cup is a reusable bell-shaped cup made from latex or silicone that is inserted into the vagina to collect menstrual fluid.

GREEN AND GORGEOUS, INSIDE AND OUT

The first step to being green and gorgeous is to look after your health from the inside. We've all heard it before—good diet and exercise will do much more for the way you look than a truckload of beauty products with miracle ingredients. The truth is that many of us think a miracle cure is easier than changing our diets, trying to fit in exercise routines and avoiding pollution. We try to drink the litre or two of water a day recommended by supermodels, but it's not always convenient. Besides which, we might not want to give up smoking, coffee, chocolate or junk food.

Get real! If you're not going to change a poor diet and lifestyle to one that's healthier, accept the fact that it's going to limit how naturally good you look and feel. The best way to produce the appearance of healthy skin is actually to have really healthy skin. No manufactured product such as foundation or concealer can do as good a job.

Once you've exhausted the possibilities of enhancing your natural beauty by looking after your health, it's time to call in the beauty reinforcements. There is much debate about how effective commercial beauty potions really are. On the one hand, some materials are very difficult for the skin to absorb, so putting them in a jar of pleasantly perfumed cream won't do much for your complexion. On the other hand, some substances can be effectively absorbed through the skin, as shown by the creams used in hormone replacement therapy and nicotine patches.

Somewhere between a simple cleansing and moisturising routine and the application of cleanser, toner, moisturiser, day cream, night cream, eye cream, lip conditioner, neck firmer, pore-refining masque and age-defying concentrate, we've crossed the line into gross over-consumption. As green consumers, can we afford to spend money and the planet's resources on a range of unnecessary items that don't really work anyway? The key is to know when to stop. By all means choose greener grooming products, but keep your beauty ritual under control.

Some occasions call for a bit of extra effort or polish, through the use of make-up, which can accentuate favourite features and disguise others. Beware of being a slave to fashion. It will cost you a fortune and take a heavier toll on the planet, just like when you're buying clothes. Apply the 'less is more' rule. Choose a small range of basic, versatile make-up colours that won't date quickly, instead of those that will be out of fashion long before the product is finished.

GREENER GROOMING PRODUCTS

So what makes a bottle of toner or a lipstick 'green'? Unfortunately, there's no definitive answer. There are no accepted standards for what constitutes an eco-friendly cosmetic, but there are ways in which cosmetics can be better or worse. Information on these issues can be hard to extract from the average manufacturer, but some cosmetic brands have embraced care for the planet and are streets ahead of conventional cosmetics.

This chapter gives you the information you need to make choices that reflect your values when looking for cosmetics and toiletries. For some this will simply mean taking the environmental record of a product or cosmetic company into account. For others, the use of animal products and testing will be the primary consideration.

When weighing up the planetary pros and cons of cosmetics, ask the following questions:
- What does the product claim to do?
- Do I believe these claims?
- Do I really need it in the first place?
- What is in the product?
- Where does it come from and how was it made?
- How environmentally friendly is its production?
- How is it tested? Is it tested on animals?
- How is it packaged?

Information that comes with the product, on its label or in its advertising, can answer some of these questions. However, in some cases it can hide the answers. Beauty is big business and the green consumer is becoming increasingly important to marketers. Learn to tell the genuine article from the greenwashed product.

COSMETICS CLAIMS

If we were to read '(1'R,6'S)-(-)-1-(2',2',6'-Trimethyl-cyclohexyl)-but-2-en-1-one' on a jar or in advertising, we probably wouldn't go near the substance, let alone put it on our skin. Yet this is the chemical name for one of the oils that gives roses their fragrance. Everything is chemical. Even water is known as H_2O. 'Natural' isn't always best, and 'synthetic' isn't always bad.

There tend to be two approaches to cosmetics marketing: I call them 'blinding with science' and 'it's only natural'.

She blinded me with science

The 'blind them with science' approach uses long chemical names, technical references and statistics. The average shopper doesn't understand scientific names, but these complicated names and scientific statements give the comforting impression that this product or substance is the very latest technological advancement in skincare and therefore must work!

When you're looking at beauty products, don't be blinded by science or afraid of it. If the manufacturers really wanted you to understand the specifics of their scientific statements, they would put them in simpler terms. If you want to know more about a product's claims, don't be afraid to ask the consultant.

Here are a few tips for deciphering the scientific claims on cosmetic labels:

- Most cosmetic companies have customer service and enquiry hotlines, so use them to clarify any marketing claims, particularly those that claim the product is better for the environment.
- Be wary of product claims that are inappropriately specific. For example, a statement like '30% younger-looking skin' tries to put a scientific and credible-sounding number on something that can't be quantified. What does it mean? Thirty per cent of what? Percentages and statistics are another way of trying to impress us with science.
- Remember that most of the people in photographs in cosmetic advertising are models, not scientists, even if they're wearing glasses and lab coats.

It's only natural!

This is the warm, fuzzy approach, where marketers avoid long chemical names and opt instead for general statements about the benefits or the ingredients. You can spot these products by the use of words such as 'pure', 'organic', 'natural', 'simple' or 'vital'. There are very few restrictions on the use of these words in product labelling. Just about anything can truthfully be said to be derived from natural sources. All ingredients used in manufacturing occur in nature first, but are mined, tapped, collected or harvested. They are then physically or chemically altered to produce products that often have no resemblance to their original source. Keep these points in mind when reading cosmetics advertising:

- Claims of being 'chemical-free' usually mean the ingredients are not synthetically derived or are considered non-hazardous.
- Don't be misled by the word 'natural'. Nature is not always kind. Just because something is natural doesn't mean that it is good for you. Many plants and animals protect themselves with chemical defence systems rather than physical strength or a protective shell. Myriad naturally occurring substances are harmful to humans. Cyanide, for example, occurs naturally in apricot kernels.
- In some cases synthetic ingredients are greener (that is, less harmful to the environment) than naturally derived ingredients. Many compounds originally found in animals, such as the fragrance from the musk deer, can be manufactured synthetically without the need to hunt and kill possibly endangered species.

How to read cosmetics labels

In Australia, there are general standards for labelling cosmetics and therapeutic goods. The primary goals of the organisations that regulate the cosmetic industry are to make sure that any products on the market are safe to use and that their ingredients lists are accurate and not misleading. However, when it comes to advertising, industry guidelines focus on claims about what the products can do, rather than how they are produced or what they are made from. The words 'natural' and 'organic' can and have been used to mean a range of different things, in a number of different contexts. Because of this there is no restriction on the use of these words in advertising. With organic produce, farmed either for food or for use in other products, it is more important to look out for 'certified organic' products, rather than the word 'organic' on its own. Ignore statements like 'earth friendly' or 'eco-conscious', unless they're backed up with specific information on exactly how the product helps the environment.

Fortunately, in addition to the advertising guidelines, there are mandatory labelling requirements for cosmetic products manufactured in, or imported into, Australia. These requirements include the listing of ingredients used to make the product. The Australian Competition and Consumer Commission (ACCC) enforces the labelling of cosmetics. This declaration of ingredients is important because it helps allergy sufferers identify and avoid exposure to problematic ingredients, it helps consumers to compare products, and it identifies the use of animal ingredients for those concerned with animal welfare.

The Australian government's National Industrial Chemicals Notification and Assessment Scheme has an online guide to cosmetics labelling at www.nicnas.gov.au.

Once a grooming product goes beyond grooming to offering a health benefit, it must be entered on the Australian Register of Therapeutic Goods, which is administered by the Therapeutic Goods Administration, before being marketed in Australia. Extensive testing and tougher product standards are required of therapeutic goods.

Product labelling, advertising and the beauty pages of glossy magazines often refer to certain ingredients or types of ingredient. Here's a list of the common groups of cosmetic ingredients, what they do and what to look for.

Emollients are moisturisers that prevent water from evaporating from the skin. Emollients moisturise, lubricate and soften the skin, prevent and ease dryness and help to protect the skin. Natural emollients tend to allow the skin to breathe better than synthetic emollients.

Look for products that contain plant oils (such as almond, avocado, coconut, grapeseed, sunflower, olive, wheatgerm and jojoba oils), cocoa butter, vegetable glycerine, lecithin and squalene (derived from plant sources). Lanolin, a wool by-product, is also an effective emollient, though vegans may wish to avoid it. Mineral oil and petrolatum (petroleum jelly) are common petroleum-derived emollients. Dimethicone and cyclomethicone are silicone emollients and should be avoided by people who are sensitive to silicones.

Emulsifiers bind the oil-based and water-based ingredients in cosmetics, holding the product together as a suspension or emulsion.

Health effects of cosmetics

In discussing the health impacts of cosmetics and their ingredients, it helps to have a broad understanding of *how* different substances can affect our bodies.

- **Allergic reactions and sensitivities** are the response of a sensitive immune system to substances that are normally harmless to most people. Symptoms of allergic reactions include skin rashes, a runny nose, itchy eyes, swelling, nausea and vomiting, and in extreme cases, anaphylaxis. It is important to note that the health impact of any given allergen will depend on the particular sensitivities of the individual who is exposed to it.
- **Toxicity** is the degree to which a given amount of a substance is poisonous, causing harm to the body's tissues. Toxins, in contrast to allergens, are poisonous to some degree to all people.
- **Carcinogenicity** refers to the way that some substances can cause or contribute to the development of cancer.
- **Oestrogenicity** refers to the way that some substances indirectly affect health by behaving in a similar way to hormones, in particular the female hormone oestrogen. Such substances are called endocrine disruptors (see chapter 2, pp. 20–1).

Many greener grooming products avoid using emulsifiers altogether by allowing the ingredients to separate but instructing the user to 'shake well before use'.

Look for products that contain lecithin (an emulsifier as well as a moisturiser), beeswax (provided you're not allergic to it) and polysorbates. Polysorbates can dry the scalp, so look for polysorbate hair products with moisturising ingredients to compensate. Many other emulsifiers can be made from either plant or petrochemical ingredients, so it's hard to list emulsifiers to avoid. Wherever possible, use products that are entirely plant-derived.

Humectants are moisturisers that attract water. They help the skin to maintain moisture. Look for vegetable glycerine or glycerol, lecithin, plant-derived glycols, sorbitol and amino acids. Avoid urea if you have sensitive skin. Urea is often used with formaldehyde, which is a suspected carcinogen and should also be avoided.

Collagen, elastin and ceramide are heavily marketed humectants and are commonly sourced from animals. There are also plant derivatives that can act like collagen (called 'pseudo–collagen'). Sun damage causes the natural collagen in human skin to deteriorate as part of the skin's ageing process. There is no evidence that collagen and elastin applied to the surface of the skin can penetrate to the deep layers of the skin where they can replace those lost through the natural ageing process.

Preservatives extend a cosmetic product's life and help to prevent the growth of bacteria, but they can also cause allergic reactions. Remember that the decaying process is part of nature and no cosmetics should last forever. The real benefits of preservatives are not to the consumer but to the manufacturer, which reduces the cost of products deteriorating on the shelf before being sold.

Look for products that contain citrus seed, grapefruit seed, rosemary and/or olive extracts, benzoic acid (which comes from many plant sources and is antifungal, but can also irritate sensitive skins) and the antioxidant vitamins A, C and E (which slow down the process that makes creams go rancid). Vitamin C is also called 'ascorbic acid', and vitamin E is also called 'tocopherol'. These ingredients help to naturally extend the shelf life of cosmetic products.

Avoid other preservatives in general. Many preservatives are antibiotic or antimicrobial agents, all of which contribute to the growth of drug-resistant 'superbugs'. Particularly avoid formaldehyde, also known as 'methanal', 'formalin' or 'formol'. Triclosan is another common antibacterial or antimicrobial agent to be avoided. Parabens are preservatives that act by killing bacteria and mould. Parabens are known skin and eye irritants, and are also suspected endocrine disruptors.

Solvents take solid or gaseous ingredients and carry them in a liquid form. They're also used in the extraction of some materials. Look for products that contain water, plant glycerol and plant-derived ethanol. Avoid methanol, acetone, turpentine, toluene, benzene, benzaldehyde and phthalates. Phthalates or phthalate esters are used as solvents in perfume and nail polish. Phthalates are suspected endocrine disruptors.

Surfactants or 'surface active agents' weaken the surface tension of water, making it penetrate the skin more easily. In creams they act as emulsifiers, while in cleansers they act as detergents. There are four main groups: sulphates, soaps, sulfonates and carboxylates.

Look for plant-derived saponins or glycosides (made from chickweed, yucca, saponaria and soapwort plants). These substances help to make a bubbly lather. Also look for amino acids and surfactants made from soy. Avoid surfactants made from mineral oil and, in particular, avoid cocamide diethanolamine, diethanolamine and triethanolamine (often abbreviated to DEA and TEA), which are skin and eye irritants and which may contain carcinogenic nitrosamines as contaminants. Also avoid sulphates such as sodium lauryl sulphate (SLS) and sodium laureth sulphate, particularly if you have sensitive skin. The toxic effects of SLS are well documented, including skin and eye irritation, the development of cataracts and other tissue damage. They should be avoided in products that are applied directly to the skin (or the scalp in the case of hair products) without dilution. They are of less concern with non-cosmetic products, such as laundry powders, in which they are diluted in water and eventually rinsed from the textile fibres. In such situations they don't come in direct concentrated contact with the skin.

Thickening and stabilising agents As the name suggests, they add body to creams and lotions. Look for products that contain carrageen and vegetable gums and waxes. Avoid mineral oil and other petroleum products.

→ **ASK TANYA**

I've heard that sodium lauryl sulphate in shampoo is carcinogenic. Is that true?

Valerie, Geelong, Victoria

Sodium lauryl sulphate (SLS) and sodium laureth sulphate (SLES) are commonly used surfactants. Claims that they are carcinogenic have begun floating around the internet and via email.

In response to public concern, the Australian government's National Industrial Chemicals Notification and Assessment Scheme (NICNAS) reviewed information from credible sources on the human health effects of SLS. Many studies confirmed that SLS is an irritant. The only carcinogenicity study found reported that SLS was not carcinogenic in beagles. In short, there is little or no evidence either way, and little reason to suspect carcinogenicity. SLES can also be an eye and skin irritant, but far less so than SLS.

The health effects of SLS are linked to the concentration and duration of exposure. Definitely avoid SLS in products that are left on the skin. Also avoid it in products (such as shampoo or liquid soap) that are applied to the skin without dilution but then rinsed off, particularly if you have sensitive skin. SLS is generally not a concern when used in household cleaning products.

Interestingly, the Cancer Council of WA notes in a 'Cancer Myths' fact sheet that the idea that SLS can cause cancer 'is a myth highly exploited by "natural products" businesses which use it to convince consumers to buy their products'.

WHAT TO LOOK FOR
General
Choose brands with an ethical, green philosophy. In the beauty industry, a company's philosophy can make a huge difference to the way it operates and the products it makes. A company driven solely by the desire to make huge profits is more likely to use cheaper, more environmentally harmful ingredients.

You may choose to look for a brand with solid environmental policies that you can trust, particularly if you don't wish to read ingredient lists in detail with every product you buy. There is a growing green market in the cosmetics industry, and more brands are choosing to factor the environment into their products, policies and practices.

For example, the Australian skincare company Jurlique has an underlying principle of 'care'—for their products, for the individual, for others, for the environment, for the land and for the planet. Its products are plant-based, not tested on animals, free from animal content and packaged in recyclable materials. Jurlique also grows many of its product ingredients on farms in South Australia. It has an organic herb farm in the Adelaide Hills, certified by the National Association for Sustainable Agriculture Australia (NASAA). Similarly, Aveda operates around environmental principles, and its manufacturing facility in Minnesota, USA, has achieved ISO 14001 environmental management certification.

Buy products with sensible, minimal packaging that can be recycled in your local area. For example, many Aesop skincare products come in simple amber glass bottles and jars that are collected by most council recycling services. Avoid over-packaged products, such as boxed fragrance gift sets.

Favour brands that use packaging that contains recycled content. For example, most of Aveda's plastic bottles are made with a minimum of 80% post-consumer recycled plastic, and the cardboard used to package make-up products is 100% recycled.

When you find a product you like, buy it in bulk. This reduces the amount of packaging used per unit of product. However, don't bulk buy cosmetics and toiletries that are likely to spoil if not used quickly.

Avoid products made with petrochemicals from the environmentally harmful petroleum industry. Look for those that state that they are 'petro-chemical-free'. Petrochemicals are also common irritants for people with allergies and chemical sensitivities.

Favour products that are entirely plant derived. However, keep in mind that plants can provide very powerful ingredients. Don't assume that all plant-derived products are suitable for sensitive skin.

Avoid palm oil unless it's from certified sustainable sources. See page 52 for more information.

Go organic. As with food, choose products made with organically grown ingredients to reduce agricultural pesticide use and pesticide residue, and to support sustainable farming practices. Look for products that state that they are certified organic. Currently, personal care products made entirely from certified organic plant ingredients are rare. Plant-based products that include some organic content are easier to find and a step in the right direction.

Look for products that are made with GM-free ingredients. Again, choose cosmetics that are free from genetically modified content for the same reasons you would choose GM-free food. For more information see chapter 7, page 93.

If you have asthma or are sensitive to inhaled scents, avoid added fragrance. The word fragrance on a product's ingredient list can cover dozens of different chemicals, many of which can trigger allergic reactions. Synthetic fragrances have been particularly implicated in inhalation allergies. If you like a little bit of fragrance, then look for products that are perfumed with plant-derived pure essential oils.

Body and skincare
Avoid aerosol body sprays and deodorants. Aerosol cans are an energy-intensive way to package and disperse liquids. Although aerosols in Australia no longer contain ozone-depleting CFCs as propellants, they may use other greenhouse gases as propellants. Aerosols are also typically a wasteful way to apply a product as they disperse a large portion into the air. Many solvents and propellants are also volatile organic compounds (VOCs), which can cause respiratory irritation and contribute to poor indoor air quality. Also, the fine mist produced by aerosols can penetrate more deeply into the lungs, increasing your ingestion of the chemical cocktail that forms the product. Try a roll-on or salt stick deodorant instead. If you must use a spray, go for a pump pack.

Avoid products containing coal tar. Coal tar is used in a range of cosmetics and toiletries, particularly in preparations used to alleviate psoriasis and eczema. Coal tar is a known carcinogen and should be avoided.

Haircare
Avoid aerosol hairsprays for the same reasons you would avoid aerosol body sprays and deodorants.

Buy shampoo and conditioner in bulk pump-dispenser packs. Shampoo and conditioner are typically products that don't spoil quickly and are therefore suited to buying in bulk to reduce packaging. We also tend to use more than we need to, often through an accidentally large squeeze of the product tube or bottle. Pump packs deliver a measured dose of product and so are great for dispensing shampoo and conditioner. Companies such as Aveda, De Lorenzo and Alchemy offer the largest sizes of their haircare products in pump packs.

Soap
Avoid antibacterial (attacks bacteria) and/or antimicrobial (attacks viruses) soaps and liquid soaps. Rather than looking after your health, they may actually do the opposite. The overuse of antibacterial products and antibiotics is leading to the growth of antibiotic-resistant bacteria. In addition, the use of these products and exposure to them weakens our immune systems, our natural and first line of defence against illness.

Buy solid soap without packaging where possible. Alternatively, buy bars of soap in multiple packs, packaged in cardboard boxes that can be recycled. Individual wrapping of soap bars is unnecessary and wasteful.

Buy 'vegetable' soap. We've said earlier to avoid petrochemical products (which includes soap) and to avoid animal-derived soap (some are made from tallow). Plant-based soaps are often labelled as 'vegetable' soap, meaning that they are plant-derived.

Make-up

Look for make-up compacts that are refillable. They are typically available for pressed powder, powder foundation and eye-shadow products. Aveda also makes a refillable lipstick case.

Go for good-quality basics instead of short-lived trend colours. As with clothing, there is a temptation to buy the current season's fashion colour, only to throw it out half-used when the next trend comes along. This wastes material resources and money.

Take mineral make-up claims with a grain of salt. 'Natural' mineral make-up, made from ground earth minerals, is a recent craze that is claiming to be healthier and environmentally friendly. As with other 'natural' claims, that doesn't necessarily means it's green. Arsenic and asbestos are both minerals. Mineral make-up products are not created equally, so you'll still have to do some research and read the labels. Ideally, they're simple minerals, like zinc oxide or iron oxide, titanium dioxide, ground up to make a powder that can be applied to the skin with a brush. It's the extra stuff that consumers should be wary of. As with other cosmetics, you still want to avoid talc, parabens, phthalates and, if you're sensitive to them, fragrances. Also look out for bismuth oxychloride, which is added for shimmer. Bismuth may be a naturally occurring mineral, but bismuth oxychloride is not. It's a by-product from smelting and also a known irritant. If you have particularly sensitive skin, it may also be helpful to choose products that are mica-free.

SUN-SAFE SKIN

Sunscreen is more important for your health than for the pursuit of beauty. Protecting our skin from the harsh rays of the sun is vital, particularly in Australia, which boasts the highest rates of skin cancer in the world.

It's important to note that using sunscreen is only one of many lines of defence we can use to reduce the harmful effects of sun exposure. Tips for reducing your exposure to the harmful wavelengths of sunlight can be found online at the SunSmart website: www.sunsmart.com.au.

Choosing a sunscreen

Sunscreens can have inorganic or organic active ingredients, or a combination of the two. Look for products that have inorganic active ingredients (usually zinc oxide or titanium dioxide) in a plant-based medium or carrier cream, rather than a petroleum-based cream or mineral oil.

Inorganic ultraviolet-screening agents act by reflecting and scattering ultraviolet (UV) radiation. Inorganic screening agents (sometimes called non-chemical sunscreens) are less likely to irritate sensitive skins. From a cosmetic point of view, micro-fine titanium dioxide products tend to look less milky on the skin than those that contain zinc oxide or zinc cream. However, health and environmental advocacy groups such as Friends of the Earth question the safety of sunscreens containing nano-sized mineral particles (that is, particles smaller than 100 nanometres, 1 nanometre being one-billionth of a metre). See nano.foe.org.au.

 Organic UV-screening chemicals work by absorbing UV light and dissipating it as heat. A range of these organic UV-screening chemicals are suspected hormone-mimicking chemicals and are therefore a potential health risk. Watch out for benzophenone-3 (Bp-3), homosalate (HMS), 4-methyl-benzylidene camphor (4-MBC), octyl-methoxycinnamate (OMC), and octyl-dimethyl-PABA (OD-PABA). For more information on these potential endocrine disruptors, see pages 20–1. Australia's Therapeutic Goods Administration hasn't advised against continued use of these chemicals. However, the Danish Environmental Protection Agency has restricted the use of 4-methyl-benzylidene camphor (4-MBC) in sunscreens.

Sunscreens with little or no petrochemical content tend to be kinder to sensitive skin and are less likely to trigger skin allergies. These plant oil ingredients are renewable, unlike mineral oils and creams, which are derived from petroleum. Sunscreen products based on plant oils include the Soleo Organics and UV Natural ranges. These alternative suncare products are often found in health-food stores.

To help protect local ecosystems, avoid wearing sunscreen when swimming in freshwater lakes and streams.

DIY GREEN GROOMING

The lowest-impact, greenest beauty products are the ones that you can make for yourself. While they may not look as pretty in your bathroom cupboard or smell as nice as commercial products, they're a lot cheaper and are worth trying. Here are some natural skincare and haircare ideas for you to try at home.

Homemade skincare and haircare
Cleansing
- Soak oatmeal in water and use the water for washing your face.
- Buttermilk can be used to gently cleanse the face. Saturate some cotton wool or the corner of a face washer with it and apply to your face. Rinse or tissue off. Buttermilk is lightly astringent and restores the acid mantle of the skin.
- Many plant oils such as almond oil can be used as a cleanser to melt away a day's build-up of make-up, dirt and pollution. Wash off with a face washer and warm water.

Exfoliation
- Bicarb soda is a great exfoliator, particularly for acne-prone skin. It softens the skin and is antiseptic. Put a dessertspoonful in your hand and make it into a paste with just water or with a little bit of cleanser. Gently rub it over your face and wash off.

Facemasks
- Mix a lightly beaten egg white with a little lemon juice to make a facemask for oily, spotty skin. Smooth it over the skin, allow it to dry, and wash off with warm water.
- Raw honey mixed with natural yoghurt makes a nourishing and soothing facemask. Apply it to damp skin, leave for up to 30 minutes, gently massage it into the skin and then rinse off.
- Make a gentle honey and oatmeal mask by putting two heaped tablespoons of oatmeal into a small saucepan with 200 millilitres of water and simmer gently for five minutes. You can also dangle a chamomile tea bag into the brew while it's simmering. Stir in a teaspoon of honey and allow the mixture to cool. When cooled sufficiently, apply the mixture to your face and leave for ten minutes. Rinse off and follow up with a moisturiser.

Toning
- Witch-hazel is an astringent, making it a good freshener and toner for oily skin. It is cheap and available from supermarkets and chemists. It is also useful as an underarm deodorant.
- Make your own almond milk toning lotion to tone and soften all skin types. Mix 4 tablespoons of almond meal with 300 millilitres of distilled water in a blender for a few minutes. Strain through calico in a funnel into a bottle.

Moisturising

- Almond and apricot oils are light enough to be used on normal or dry skins as a facial moisturiser.

Hair colouring

- Steep rosemary leaves in water and use the water as a hair rinse to enhance the colour of dark brown hair. Chamomile flowers can be similarly used for blonde hair. Water-diluted lemon juice can be used to highlight brown hair.

Body and bath

- Almond, jojoba and even olive oils are great moisturisers for the skin. Rub a little into your skin, especially your legs, knees and heels. Almond oil is light and well absorbed into the skin.
- Wheatgerm oil is a particularly good moisturising oil and can help to improve the skin's elasticity. For this reason it is great to rub into expanding pregnant bellies (mix in a few drops of lavender and neroli oil). It is also rich in vitamin E.
- Aromatherapy oils can be used instead of perfume. Mix your own blends, using your favourites. However, some oils shouldn't be applied directly to the skin or used during pregnancy. Instead, dab a little onto your clothes or onto a tissue or hanky worn underneath your clothing. Aromatherapy oils can also be added to your bath.
- Add some bicarb soda to your bath to soften your skin.
- A bucket of warm water with a knob of grated ginger and a few drops of tea-tree and peppermint oils makes a great foot soak.

Miscellaneous

- Soothe puffy eyes with slices of cucumber or cold, wet tea bags.
- Aloe vera soothes irritated skin and helps to heal sunburn. Cut the leaf and use the clear gel inside.
- Floral waters are inexpensive, natural alternatives to perfumes and fresheners.

→ ASK TANYA

My question is about hair removal. There are plenty of eco-friendly shampoos, conditioners and facial creams etc. out there. However, I was wondering which form of hair removal (waxing or shaving etc.) has the least effects on the environment.

Li Yen, Willetton, WA

I've always said that for me there's a time and a place for old-growth deforestation: the time is spring, and the place is below my knees. Allowing yourself to be naturally hairy is the greenest option, but I recognise that free-range armpits are not everyone's idea of attractive. The impacts of the various hair-removal options are many, diverse and relatively minor, and weighing them up can be (dare I say it) splitting hairs!

Looking at the home DIY options, you can either remove hair above the surface of the skin by shaving or using a chemical depilatory, or the entire hair (root and all) by waxing, sugaring, tweezing or using an electric epilation device. The first method lasts a few days, the second a few weeks.

Shaving is less green when done in the shower (using 7–20 litres of water per minute) or under a running tap with disposable razors and aerosol shaving creams. Aerosols are a relatively energy-intensive form of packaging and should be limited. If shaving is your preference because it's less painful, use longer-lasting shavers, perhaps with replaceable heads, and clean and dry them properly to help the blade last longer. Use a basin of warm water, instead of extra time in the shower.

Chemical depilatory products have been called 'chemical shaving'. They commonly use calcium thioglycolate, which breaks down protein in the hair it comes into contact with, allowing hair above the skin to be scraped or wiped away. While not a major environmental concern, calcium thioglycolate can irritate the skin and tends to smell awful. Like shaving, chemical depilation needs to be done frequently, increasing your total consumption of this product.

The various methods of yanking hairs out hurt a bit, but they result in more hair-free days before regrowth appears. Electric epilation gadgets, such as the 'Emjoi Gently' (I call it the 'tweezerama'), have multiple 'tweezer discs' attached to a rotating drum. They are less messy than waxing or sugaring but use a small amount of electricity, which, along with all your electricity needs, can be supplied by accredited GreenPower (www.greenpower.gov.au).

Wax, used in conjunction with disposable strips, is made from beeswax, plant waxes and/or petrochemically derived waxes (such as paraffin). The extraction and production of petrochemicals is generally polluting, though plant-derived alternatives are not without their own environmental impacts. Vegans may wish to avoid beeswax products. Wax is hard to wash away, so applicators and strips tend to be thrown out, adding to the waste you send to landfill.

Sugaring is an alternative to waxing, using a syrupy sugar-based mixture instead of hot wax. You can buy a few sugar-based 'natural' products or sugar waxing kits, or you can make your own. I did myself by simmering 1 cup of sugar with 1 cup of honey, the juice of half a lemon and several drops of tea-tree oil (note for vegans: there are other recipes that don't use honey—Google 'sugaring'). I cut up an old flannelette nappy to use as strips. Traditional wax grips the skin more and so hurts more. Sugaring is less painful, cheaper if homemade and, being water soluble, is far easier to clean up after than wax. The cloth strips can be washed and reused.

ANIMAL WELFARE ISSUES

Some pretty nasty things have been done to or with animals to produce cosmetics for humans. Cosmetic companies are constantly developing new products to keep ahead of their competition in the highly lucrative beauty market. In an effort to fulfil legal health and safety requirements, the products have to be tested, often on guinea pigs and other animals. These test animals often live in constant pain and in pitiful conditions. Animal testing and use of animal ingredients in cosmetics is particularly controversial because these products are non-essential, luxury items. Animal rights advocates understandably question whether these animals should be made to suffer and die for the sake of human vanity, particularly when there are alternatives.

Animal testing

The truth is that virtually all the ingredients in cosmetics and toiletries have at some time been tested on animals but, until recently, most of us were unaware of the suffering this entailed. Fortunately, a few companies concerned with animal welfare issues offered us alternative products that were not tested on animals and supported the development of alternative testing methods.

Find out about a brand's testing policy before buying it. Make sure that the manufacturer extends its policy to its suppliers. For example, it has been The Body Shop's policy not to buy any ingredient tested on animals by their suppliers for cosmetic purposes after 31 December 1990. The program required suppliers to submit a biannual declaration on every ingredient supplied, and its supplier monitoring systems were independently audited against the international quality assurance standard ISO 9002. Other cosmetic companies, including Avon, have said that consumer behaviour, boycotts and demonstrations were major influences in their decision to stop animal testing.

Note that the words 'natural', 'herbal' and 'organic' on packaging and in advertising do not in any way relate to whether the product is cruelty-free or not. Be wary of cruelty-free claims. Some companies that make such claims actually commission other laboratories to do their testing for them. Animal welfare consumer group Choose Cruelty Free (CCF) has assembled and regularly updates records of reliable cruelty-free brands and products. Their preferred products list includes only brands that have applied for and received accreditation from CCF that the company is not using animal testing. The preferred products list also indicates brands that make only vegan products (containing no animal content whatsoever) or some vegan products. Go to www.choosecrueltyfree.org.au.

Animal ingredients

'Cruelty-free' in the language of the beauty and toiletries industries refers only to animal testing and not to the use of animal ingredients. Many animal ingredients are by-products from other industries. It can be argued from an environmental point of view that these products make use of animal by-products that would otherwise contribute to the amount of waste going to landfill. More worrying are animal-derived cosmetics ingredients specifically for which the animal is killed. If the use of animal by-products concerns you, here is a list of common animal by-products used in cosmetics manufacture, which you can check against a product's ingredients list.

- **Ambergris** is taken from the intestinal tract of the sperm whale (or alternatively its vomit) and used as a fixative in perfumes. Whale ingredients, including ambergris and spermaceti, are no longer used in Australia, Europe and North America. However, some countries still allow them, so certain products from such countries may contain them.

- **Castoreum** is a secretion from the beaver's scent glands, used as a fixative in perfumes.
- **Collagen** is a protein found in connective tissue, skin and bones, which forms gelatine on boiling. There are claims that it can counteract the effects of ageing and overexposure in human skin, but there is much debate between medical experts over its effectiveness. Its molecules are believed to be too large to enter the skin.
- **Elastin** is used to improve the elasticity of skin. It is often made from animal sources.
- **Glycerine** (also glycerol or glycerin) is a thick, sweet, syrupy liquid used as a solvent, emollient or lubricant in skin creams. It can be derived from animal fat as a by-product of soap manufacture, or produced synthetically from propylene alcohol or naturally from the coconut palm or other vegetable oils. Currently, manufacturers do not have to list whether the glycerine used is animal derived or plant derived. Fortunately, it is usually plant derived.
- **Hyaluronic acid** is derived from the combs of roosters and is used in skincare products.
- **Mink oil** is a by-product of the mink fur industry and is used in shampoos, conditioners and as an emollient in moisturisers.
- **Musk** is a dried secretion from glands of the northern Asian hornless deer. It is used in perfumes, although most manufacturers now use a synthetic version.
- **Oestrogen** is a hormone used in skincare. Oestrogen can be made synthetically but animal oestrogen is cheaper to produce.
- **Spermaceti** is a wax made from sperm whale tissue and is used in some creams and shampoos. Also see ambergris.
- **Stearic acid** is a fatty acid used in a large number of cosmetics to give pearliness to lotions. It can be obtained from both plant and animal sources. Legally, the ingredients do not have to state the source, but some manufacturers now state whether it is obtained from palm oil or tallow.
- **Tallow** is an animal fat obtained by boiling the organs and tissues of sheep and cattle. It is used in soap, lipstick, shampoos and shaving creams.

MAKE A COMMITMENT

This week I will...

☐ _____

☐ _____

☐ _____

This month I will...

☐ _____

☐ _____

☐ _____

When I get the opportunity I will...

☐ _____

☐ _____

☐ _____

Notes

_____

_____

_____

_____

_____

_____

7

Food

Food provides the energy and nutrients that are the building blocks for healthy bodies. Yet we tend to over-consume food, just as we over-consume natural resources. More than half the Australian population is overweight or obese, a common side effect of 'affluenza'. Just as we need to find balance in the way we live environmentally, we also need to find balance in the way we eat and the foods we choose. Sustainable food choices involve looking after the health of the inner environment of our bodies as well as the outer environment of the planet.

All our own food-related activities and those that bring food to shops and restaurants have an impact on the environment. For example, a kid's meal from a fast-food chain has a far greater toll on the environment than a cheese and salad sandwich made from local produce. Every bite we eat comes at an environmental cost, but it also represents a chance to be healthy. Being green in the kitchen is about looking after the inner environment of our bodies as well as the outer environment of the planet.

HOW FOOD IS PRODUCED

Before we look at how we can be greener with the food we eat, we need to understand where it comes from. Nine out of ten Australians now live in urban areas. It's easy to lose touch with the land and the way food is made when you're surrounded by high-rises and the traffic of the CBD, or by the brick veneers and gardens of suburbia. Those with a vegie patch may have more of an idea, but even that won't give a full appreciation of the giant task of feeding Australia's populations and the associated environmental implications.

Take, for example, the humble loaf of bread. Most people looking at the loaf would say that the only environmental problem with a bag of white sliced bread is the bag itself, because it's not biodegradable and is even more of a problem if it's littered. In fact, that loaf of bread represents an investment of many environmental resources. It took topsoil and water to grow the wheat, energy to grind the wheat into flour, more energy and water to make it into a loaf, and fuel to deliver it.

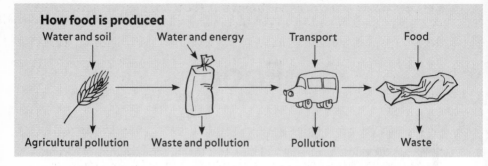

How food is produced

Water and soil	Water and energy	Transport	Food
Agricultural pollution	Waste and pollution	Pollution	Waste

Similarly land, water and energy are used to produce the food in our local shops. The fresh produce and simple staples and ingredients require resources, and processed foods require more again. It's a complex system, so it's hard to make a claim that an apple is 'green' while another food isn't. That's why there are no overarching environmental labels for food. There are some that focus on a single issue. For example, the Carbon Reduction Label tells you something about the greenhouse impact of a food but doesn't cover the pesticides or fertilisers used to grow the food. Certified organic labels tell you something about these chemical inputs, but the food can be packaged or further processed in all manner of ways. Just as the Heart Foundation Tick is important for someone concerned about weight gain or heart disease but is not an indicator of a 'safe' food for an allergy sufferer, it's unlikely that there will ever be a simple overarching 'green tick' for food. Food labels and eco-labels can provide information, but they can't do your thinking for you.

FOOD TIPS

The tips in the coming pages are things to keep in mind when making food choices. Your choices might be different from someone else's, depending on what values are closest to your heart. For some people, animal rights and welfare are the priority; for others, it may be organic farming. There's a lot to consider, but fortunately some underlying themes start to emerge. Food sustainability is really based on common sense: not taking more than your fair share, looking for variety, preserving biodiversity, avoiding waste and making healthy choices.

$ **Don't waste food**. We can go a long way towards stopping food waste by planning meals and writing a shopping list before buying groceries, serving portions that match our appetite, resisting the urge to 'supersize' takeaway meals for the sake of a cheaper meal deal, and by saving and freezing leftovers for later meals and snacks. Food waste is covered in detail in chapter 8, page 105.

Get your vitamins and minerals from fresh food, not supplements. Some food additives, such as iron supplements for those with anaemia, are good for us. However, they are generally better absorbed from sources in which they occur naturally and require fewer resources to be produced. Our bodies are designed to absorb nutrients from complex foods, not concentrated pills, so never assume that supplements can replace a balanced diet.

Go organic. It's worth going organic for the good of your health, as well as that of farmers and the environment. Organic farming produces food without the use of artificial pesticides or fertilisers. A mix of crops (known as a 'polyculture') are grown, instead of a single crop (a 'monoculture'), which can deplete the soil. Many organic farmers keep livestock alongside their plant crops. The manure provides natural fertiliser to keep the soil healthy and productive. Note that trace amounts of residue faecal matter can carry pathogens, so wash organic produce, just as you'd wash conventionally grown produce. Don't be put off by this—it's nature's recycling system.

Food crops are rotated with soil-enriching plant crops such as legumes, which replenish the nitrates in the soil. Organic food contains no artificial fertilisers, pesticides, hormones, growth stimulants, antibiotics, added waxes or finishes or other added synthetic chemicals. It is also free from genetically modified content and is not irradiated. For consumers, this means that they are putting fewer pollutants into their bodies and into the environment.

Organic food is now available in many larger supermarkets. As well as fresh produce, there are ranges of canned organic food, organic pasta, organic dairy and soy products, and baby food. You can also look for restaurants that offer food prepared with organic ingredients, as well as those with organic wines on the wine list.

Make sure it's really organic. There are few restrictions on the use of the word 'organic'. Look for products that state 'certified organic' on their labels. This means that their farming and production methods have met the criteria of independent organic certifying bodies.

Avoid genetically modified (GM) foods. Foods that contain GM ingredients are required by law to be labelled with the words 'genetically modified'. There may be some contamination of up to 1% GM content per ingredient before this 'unintentional' inclusion of GM content is required to be labelled. The main GM ingredients on Australian supermarket shelves are derived from imported canola, corn (or maize), tomato and soy products, and both local and imported cottonseed products. Other crops that have been modified overseas include wheat, rice, chicory, squash, potato, alfalfa and cotton.

GM foods are already out there on the supermarket shelves, and the people who eat them are unwittingly part of an unofficial field trial. If you don't want to be part of the GM food experiment and would rather take a precautionary approach, look for foods labelled 'GM-free', 'not genetically modified' or 'certified organic'. Greenpeace Australia has produced the *True Food Guide* to help consumers identify GM-free brands. For more information, go to www.truefood.org.au.

Avoid unnecessary packaging. While packaging helps preserve some foods, it can also serve no other purpose than to carry branding and advertising. Choose unpackaged or sensibly packaged foods, as outlined on page 108.

Meat and poultry

Eat less meat to use less land, reduce emissions and use less water. Try to replace one meat meal each week with a vegetarian alternative, such as a Chinese tofu stir-fry or Mexican bean tacos and burritos. In Western societies we eat far more meat, poultry and fish products than we need to, given their nutritional value. This taste for meat, along with fatty takeaway foods and general overeating, is making these societies increasingly overweight and unhealthy, particularly in the USA and Australia.

You would expect the world's grain crops to be grown to produce food for people to eat. This isn't always so. A lot of the grain grown is used to feed animals on ranches and intensive livestock farms, which in turn are used for meat and other animal food products. Animal rights considerations aside, you can produce more food with a given patch of land by growing plant foods than you can by grazing cattle or by producing food to rear food animals. A significant amount of the land cleared in South America is intended to provide more grazing land for cattle that will one day end up as beef patties, feeding America's supersized love of the burger meal.

Even if we put animal rights issues to one side, we should question the amount of meat, particularly beef, we consume and look at other protein sources. The hoofs of livestock are wreaking havoc on Australia's thin layer of topsoil. In addition, livestock are a significant source of greenhouse gas emissions, and it takes a huge amount of water (see page 100) to produce animal protein.

How far can a grain go?

Not all the grain grown in the world is turned directly into food for humans. Some of it becomes livestock feed to produce other foods. For example, grain is fed to cattle to produce beef.

According to the Worldwatch Institute it takes:

- 7 kilograms of grain to produce 1 kilogram of beef
- 4 kilograms of grain to produce 1 kilogram of pork
- just over 2 kilograms of grain to produce 1 kilogram of poultry
- less than 2 kilograms of grain to produce 1 kilogram of farmed fish.

Buy free-range and organic animal products. Don't forget that there are also alternatives to conventionally farmed animal products. Buy free-range, vegetarian eggs instead of those laid by battery hens fed with feed that contains antibiotics and animal content. Free-range chickens and organic meat are available from some butchers, supermarkets and health-food stores. For pork products, note that 'free-range bred' is not the same as 'free-range'. 'Free-range bred' means that the pork comes from pigs that were born and raised with access to outdoor areas (the general meaning of 'free-range'), but are subsequently raised indoors, typically transferred at the age of about three weeks. For the next several months they are raised intensively indoors.

Avoid over-packaged meats. Meat and poultry are often sold in supermarkets on polystyrene or other plastic meat trays. This is an unnecessary use of materials. Polystyrene is rarely recycled. Buy meat in simple plastic bags or bring your own reusable containers.

→ ASK TANYA

I have been reading about the labelling that we find on egg cartons. I have now found a label that is 'hencoop'. The eggs are organic grain-fed hencoop eggs. Does that mean the eggs come from hens that are fed organic grain but are not free-range?

Sonia

I think you're right. Organic feed alone does not make chickens or eggs organic, as organic standards include specifications for the living conditions of the hens. The use of the word 'hencoop' sounds like creative language to me, intended to conjure up images of picturesque storybook farms. Look for 'certified organic' eggs, certified by a reputable organisation.

Seafood
Buy seafood from sustainable sources and avoid overfished species. Seafood is becoming an increasingly popular protein food, particularly among people who wish to avoid red meat. Many have also heard about the health benefits of omega-3 fatty acids, commonly found in cold-water fish. However, the increasing demand for seafood is causing the depletion of fish stocks around the world. Sustainable seafood is harvested in limited quantities from carefully managed stocks. This ensures that individual species aren't overfished, marine habitat is preserved and pollution is prevented.

The Australian Marine Conservation Society (AMCS) has produced a national guide to choosing sustainable seafood. It uses a traffic light rating system: green means the seafood is a 'better choice'; orange indicates 'think twice'; and red means 'say no'. The *Sustainable Seafood Guide* can be ordered online at www.amcs.org.au. Use this guide to make more informed choices about buying seafood.

There is also a great website at www.goodfishbadfish.com.au that features a 'Seafood Converter'. Select the fish you're considering from a drop-down menu and the converter will show you the AMCS traffic light rating and other names the fish may be sold as, and if it's not one of the better choices, you'll also get a list of up to three more sustainable alternatives.

Also look out for seafood products bearing the Marine Stewardship Council (MSC) label. The Marine Stewardship Council is an independent, global, non-profit organisation that recognises sustainably managed fisheries through a

certification program. Products bearing their logo have been approved by the MSC program. Their website also lists the certified products available in the countries where they are active, including Australia and New Zealand. Visit www.msc.org.

Avoid fish with high mercury levels. Mercury is highly toxic to humans. Industrial and marine pollution has released mercury into many lakes, rivers and oceans, where it accumulates in the tissues of marine animals. Mercury tends to concentrate or 'bioaccumulate' in the food chain, so larger predatory fish tend to have high levels of mercury in their flesh. Fish that typically have high levels of mercury include shark (often sold as 'flake'), ray, swordfish, barramundi, gemfish, orange roughy, ling and southern bluefin tuna. Limit your intake of fish of these species. Pregnant women, women of childbearing age and young children are advised to limit their consumption to one meal a month.

Canned tuna is not included in this list of fish to avoid because smaller, younger tuna are used in canning and are naturally lower in mercury, not having lived long enough to accumulate dangerously high levels of this heavy metal.

Avoid shark's fin products. Shark's fin soup is a traditional Chinese dish, considered a delicacy. While the fins of sharks fetch high prices, the market for the rest of the shark is much less lucrative. In some parts of the world it has become common practice for fishermen to cut the fins off live sharks and dump the dead or dying carcasses back into the ocean, a practice illegal in Australian waters. This is a highly wasteful and inhumane practice, but one difficult to police. Many wild shark populations are under threat because of shark finning, overhunting and other pressures. Not all shark fin products are caught this way. However, it is impossible to tell which is which. Avoid eating dishes prepared with shark's fin to be confident that you are not supporting illegal shark finning.

Choose canned seafood products that avoid harming other marine species. Some fishing practices, such as the use of drift nets or gillnets, catch aquatic life indiscriminately. Non-target species or 'by-catch' include sea turtles, dolphins and dugongs. These animals can become entangled in nets or lines and can be seriously injured or killed. Fisheries can instead use other fishing methods that minimise by-catch. Look for 'dolphin-friendly' canned tuna and similar seafood products that use less harmful fishing methods. In particular, favour MSC-certified products.

Fresh produce
Buy fresh local produce in season. Advances in harvesting, preserving and packaging fruit and vegetables have meant that they don't have to be consumed locally. Many varieties of fruit are picked before they are ripe so that they can be transported over long distances, ripening along the way. Harvesting and transportation are carefully timed so that the fruit looks its best when it's finally on display in the shop. Huge amounts of fuel are used to transport these foods great distances.

Buy locally grown fruit and vegetables. Picked in season, they taste better and are often better for you. Buying them also supports Australia's farmers. Buying local produce in season will reduce the range of choice, but you can still enjoy variety by using these foods in different ways. The change in availability of produce through the year is part of what makes the seasons interesting.

Support your local farmers' market. Farmers' markets support local farmers and cut out the middle man. In addition, they offer local produce in season.

See if you can grow your own. Fruit trees and a backyard vegie patch can provide you with some fresh food and herbs. If you have children and live in the city, growing your own food can help your children to learn where food comes from, providing an important environmental lesson. Water your vegie patch with water from a rainwater tank so that you don't add to the amount of water you take from urban water mains.

Processed and packaged food
Eat less processed food. Processed food has had a lot of changes made to it. The apricots in the fresh produce section, for example, are different from

those in the canned fruit aisle and vastly different from those in a jar of apricot jam. Processing food involves pre-cooking the food, chopping it up or adding extra ingredients or preservatives. Processing usually requires more energy and water use, so has an additional cost to the environment. Some of the fibre (and goodness) may be removed from grains, such as rice and wheat, to make them whiter and softer. Processing in some cases means that the food can last longer or is easier to use. Unfortunately, some of the nutritional value of food can be lost when it's processed.

Food additives to avoid and their code numbers

- **Artificial sweeteners:** aspartame (951) and saccharin (954)—thought to be potentially carcinogenic
- **Glutamates:** (620–635), particularly monosodium glutamate or MSG (621)—a flavour enhancer that some people are particularly sensitive to
- **Sulphites:** (220–228)—can cause severe asthma attacks and other breathing difficulties in sulphite-sensitive asthmatics

'E numbers' (the code numbers commonly used in ingredient lists) were developed by the European Community for the declaration of foodstuff additives and are used in many countries around the world. For a list of the food additives approved by the European Union and their numbers, visit the Federation of European Food Additives and Food Enzymes Industries website at www.elc-eu.org.

Avoid products with a lot of additives. Look at the ingredient list on the pack and avoid products with a long list of additives, artificial colours and flavours, particularly if you have food sensitivities or allergies. Colouring is added to foods to make them look more appetising. Flour and other products are sometimes bleached to make them whiter. Preservatives are added to foods to give them a longer shelf life. Special additives can also keep oil and water mixed in a salad dressing, or keep powdered spices from clumping together. A host of additives are put into foods to make them smell nicer and taste stronger, to change their colour and artificially increase their nutritional value. Additives can also cause allergic reactions, make food harder to digest, be bad for your health and even cause hyperactivity in children (also known as the 'red cordial' effect).

Beverages

Use a water filter. Remove impurities and added chemicals from your drinking water by using a water filter or purifier. Make sure you change the cartridge regularly. Use filtered water in a refillable drink bottle instead of buying bottled water.

Buy loose tea leaves instead of tea bags. Tea bags can make tea-making more convenient. However, the bag, string and tag require extra resources in their production, and they just become more waste to dispose of once you've made the tea. Particularly avoid tea bags that are individually wrapped.

Don't buy coffee bags. Coffee bags are coffee grounds that come in porous bags like tea bags. They are generally wrapped to 'seal in freshness' and prevent the grounds from going stale. When you count the box and the cellophane wrapper, these products put four layers of packaging between you and your coffee.

Go for organic, Fairtrade hot drinks. As well as organically grown varieties, there are also a number of Fairtrade tea, coffee and hot chocolate products available. Look for Fairtrade coffee in particular. Much of our coffee comes from poor, small-scale farmers in developing countries, who see as little as 3 cents from your $3 morning latte. Pay a little more for Fairtrade coffee to ensure that farmers get a fairer deal.

Bottled water boom

Australians now spend about half a billion dollars on bottled water. Most of the bottles are made from PET—a petroleum-based plastic. Although PET is recyclable, only a small percentage of the global consumption of these bottles is recycled. In Australia, our PET recycling rate is about 36%, but that still means a lot of bottles are being wasted. In short, bottled water is a wasteful way to quench your thirst.

FOOD'S CARBON FOOTPRINT

Producing food also has greenhouse implications. Aside from greenhouse emissions from agriculture itself, the processing and transport of food requires energy, the bulk of which comes from burning fossil fuels. The more heavily processed a food product is and the further it has to be transported, the higher its energy cost and global warming contribution.

The Carbon Trust, a British government–backed non-profit organisation, developed the Carbon Reduction Label as a way for producers and manufacturers to measure the total carbon impact of their products—from raw materials, manufacture and distribution to consumer use and disposal or recycling of packaging—and make a commitment to reduce it by a certain amount. The label is a form of disclosure, standardising the way that brands can declare their carbon footprint and how they're reducing it. Environment group Planet Ark has partnered with Carbon Trust to administer the program in Australia. For more information, visit www.carbonreductionlabel.com.au.

Food miles

'Food miles' are another hot topic relating to the carbon footprint of food. They are simply a measure of the distance a food has come from farm to fork, with the idea being that the further it has been transported, the higher its energy needs and the greater the carbon impact. However, this is an over-simplification. This measure does not take into account things like the influence of refrigerated transport, which has higher energy costs, or the limits of growing food locally. In cold climates, warm-climate foods can be grown in climate-controlled greenhouses with much higher greenhouse emissions but low food miles. Consideration of food miles alone will not necessarily result in the greenest or most ethical food choices. For example, strict observance of a low food mile diet would prevent people from supporting Fairtrade farming initiatives in developing countries overseas. In short, food miles are something to keep in mind, but they're not the be-all and end-all of food sustainability.

FOOD'S WATER FOOTPRINT

Another issue getting a lot of press is the water footprint or virtual water content of food. Secure supplies of fresh water for agriculture, industry and households are becoming an increasingly hot topic as the world starts to feel the effects of climate change, drought and water pollution. Virtual water, also known as 'embodied water', is defined as the total amount of water required to produce a commodity or service. This includes soil moisture and natural rainfall, not just irrigation.

Householders need to start thinking about virtual water. For the green consumer, this primarily means avoiding food waste to avoid wasting the

Virtual water content of various foods*

ITEM	VIRTUAL WATER CONTENT (LITRES)
1 glass of beer (250 ml)	75
1 glass of milk (200 ml)	200
1 cup of espresso coffee (125 ml)	140
1 cup of tea (250 ml)	35
1 glass of wine (125 ml)	120
1 slice of bread	40
1 apple	70
1 glass of apple juice (200 ml)	190
2 potatoes (200 g)	50
1 bag of potato crisps (200 g)	185
1 egg (40 g)	135
1 beef burger (150 g)	2400
1 tomato	13

• Global average, recognising that some of these foods are imported

water that was required to produce it. We can also reduce our consumption of foods with a high water cost and choose foods that are less water-intensive, but we can also buy foods with a high virtual water content that are grown in wet climates. There's a potential conflict here with the idea of buying local produce: we also need to recognise that there are some foods that can't and shouldn't be grown in Australia's climate and conditions, and these foods, used sparingly, are perhaps more appropriately imported. Keep in mind the virtual water content of food when shopping, and use wisely those that have a high water cost. For more information, visit www.waterfootprint.org.

MAKE A COMMITMENT

This week I will...

☐ _____

☐ _____

☐ _____

This month I will...

☐ _____

☐ _____

☐ _____

When I get the opportunity I will...

☐ _____

☐ _____

☐ _____

8

Waste and recycling

Waste is simply unwanted, unusable or inconvenient by-products—the extra stuff produced by human activities for which we don't have further use. We're the only species that produces waste that other species can't make use of. Large animals produce waste in the form of dung, which is a feast for dung beetles and fertiliser for the soil. When a sea creature gets rid of its shell, a hermit crab moves in or a small octopus uses it as a temporary hiding place.

Nature recycles and purifies water through the water cycle of evaporation, transpiration and precipitation; carbon is recycled through the growth and decay of living things. In contrast to nature's cyclic flow of materials, humans alone have come up with a linear flow of materials from mining or harvesting raw materials, to manufacturing, to product use and finally to the dead end of disposal.

WHY REDUCE WASTE AND RECYCLE?
When you look inside the average rubbish bin, you see (and smell) rotting food, empty packaging, bits of plastic and metal, and yesterday's news and celebrity gossip. What you don't see is the water and topsoil required to grow that food, the 170 megajoules per kilogram of energy needed to make that bit of foil or aluminium drink can, or the piece of land needed to grow the trees that became glossy magazines. There's much more to developing greener habits with waste than recycling the odd bottle or can. It's all about considering our material resources, nature's ability to keep providing them to a growing global population, and how we can make much better use of them. It's about using materials wisely, keeping them in circulation and out of landfill.

We should reduce our production of waste and recycle what we can for these reasons:
- **to make better use of material resources**. Mineral resources, such as metals, are non-renewable. By making better use of what has already been mined, we can avoid the high costs of scarcity and the need to further damage ecosystems by extracting more of these materials

- **to save energy**. Making new products from recycled materials often requires less energy than using raw materials. This reduced energy use also means less greenhouse gases emitted
- **to reduce greenhouse emissions**. As well as the reduced emissions from saved energy, rotting waste in landfill produces greenhouse gases, such as methane
- **to recover potentially useful materials**. One person's trash is another's treasure. It seems a waste to send to landfill a drink can that could be recycled again and again, or food waste that could be composted, transformed into plant food and turned into food again in a backyard vegetable patch
- **to reduce the environmental costs of waste collection and landfill**. Though this is not the number-one reason for reducing waste in Australia, it's important to keep in mind that garbage and recycling collection requires petrol to power the trucks, and landfills and waste-management sites need land a reasonable distance from population centres.

Is biodegradability the answer?

Biodegradation is the breakdown of organic (living or once living) materials or the products made from them. It happens through the action of worms, insects and micro-organisms. Substances are said to be biodegradable if they decompose naturally into simpler materials, instead of hanging around in the environment. It's nature's recycling system of breaking down old materials into the simple substances and nutrients that are the building blocks for new living materials.

One aspect of biodegradability is the breakdown of unwanted solid materials. This may be rubbish sent to landfill tips or litter polluting urban areas of the natural environment. Materials made from plants or animal ingredients, such as paper, leftover food or a cotton shirt, are biodegradable. In contrast, the microbes that normally munch through matter don't know what to do with plastics. As a result, plastics can sit in landfill for years—even centuries—and not break down. This can be a problem in countries such as Japan, Italy and the United Kingdom, where there is a shortage of landfill space.

The level of heat, light, water and oxygen and other environmental conditions can speed up or slow down biodegradation and change the way the chemical processes happen. For example, when organic materials break down after being buried in landfill, oxygen and sunlight are blocked out and more methane gas is produced. This is a problem because methane is a greenhouse gas and an explosion risk in high concentrations, and it doesn't smell too good! It's far better avoid producing waste, even biodegradable waste, in the first place, and recycle the biodegradable waste through composting, Bokashi bins, and paper and green waste recycling collections, assuming your council provides them.

WHAT'S IN OUR WASTE?

As with saving energy and water, a smart place to start acting on waste is by looking at a snapshot of the waste we're producing. A while ago a study was done of the contents of the rubbish bins of homes around Australia. It found that the average household throws away over 15 kilograms of rubbish per week. Good waste management involves trying to produce less waste in total, recycling and reusing what you can and throwing it away in the right bin. Here is a look at what they found in our bins, how much of each type they found and how these waste items might otherwise be dealt with:

You'll notice that, in addition to the recyclable items, the list identifies items that are avoidable or those that can be given to charity collections, as they're reusable.

What's in our bins?

Paper and cardboard:
23.6 per cent (generally recyclable)

Food waste:
23.5 per cent (avoidable and/or compostable)

Green and garden waste:
20.2 per cent (compostable and/or recyclable through green waste collections)

Other: 12.3 per cent (clothing and household items can be given to charity)

Glass, such as bottles and jars:
8.4 per cent (recyclable)

Plastics: 6.1 per cent (mostly recyclable)

Contamination: 2.6 per cent (landfill)

Steel, such as food cans:
2.4 per cent (recyclable)

Liquid paperboard, such as milk cartons:
0.5 per cent (recyclable)

Aluminium, such as drink cans:
0.3 per cent (recyclable)

Hazardous: 0.2 per cent (special collections).

Source: based on information from the Beverage Industry Environment Council

FOOD WASTE

One of the biggest contributors to household waste is food waste. Some food waste is unavoidable, such as the inedible or unpalatable portions of fruit and vegetables, like the green bits on potatoes. Some food waste is avoidable, like the dairy food that you forgot about until after it had reached its use-by date. All food waste sent to landfill, despite being biodegradable, represents a waste of resources and a contribution to greenhouse emissions. The good news is that 'reduce, reuse, recycle' applies as well to food waste as it does to bottles and cans. We can start by reducing the amount of food that goes to waste, 'reusing' leftovers where it's not unhealthy to do so, and recycling food waste into plant food by composting.

Reducing food waste

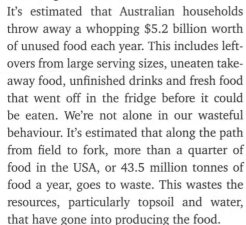 It's estimated that Australian households throw away a whopping $5.2 billion worth of unused food each year. This includes leftovers from large serving sizes, uneaten takeaway food, unfinished drinks and fresh food that went off in the fridge before it could be eaten. We're not alone in our wasteful behaviour. It's estimated that along the path from field to fork, more than a quarter of food in the USA, or 43.5 million tonnes of food a year, goes to waste. This wastes the resources, particularly topsoil and water, that have gone into producing the food.

Here are some tips for avoiding food waste:

- Write a meal plan for the next few days, and write a shopping list based on this. This will help you to avoid buying perishable foods that might go off before you get around to using them.
- Remember that some fresh produce, such as pumpkin, citrus and potatoes, last longer than things like leafy salad greens. If you eat a lot of salads, you may have to shop a little more frequently.
- Store food appropriately. For example, refrigerate perishable foods, and store garlic and onions in a cool, dark and dry

Garbage or sewage?

Let's get something straight: in urban areas, the toilet and the sink connect to the sewage system, which is designed to take care of wastewater, urine and faecal matter. There is an entirely separate system to collect rubbish and recyclable items. Toilets and sinks are not meant to be garbage disposal units. Food scraps that can't be composted, such as meat, should be put in a Bokashi bucket or the garbage. In-sink garbage disposal units (or food scraps flushed down the toilet) only add more solid material to wastewater, increasing the load of the water treatment process. Either way, the waste will produce methane, whether it's at the sewage plant or a landfill tip. If the tip or sewage plant captures methane and turns it into electricity, that's a bonus. However, in-sink garbage disposal units should be avoided as they are an unnecessary and very water-thirsty device.

Equally, the toilet is not a rubbish bin. Cotton buds, used nappies and feminine hygiene products should not be flushed as they make the job of sewage treatment all the more difficult and can cause plumbing blockages.

place. Also store potatoes in a cool, dark place, but separate from onions, as the two give off gases that cause each other to spoil more quickly. The New South Wales government's 'Love Food, Hate Waste' website has more tips on storing a variety of foods—visit www.lovefoodhatewaste.nsw.gov.au.

- Only put on a plate what you know you will eat. You can always go back for seconds. Freeze or refrigerate the leftovers, and reheat them for tomorrow's lunch.
- Resist the temptation to 'supersize' fast-food meals. They are not good value if you don't eat them.
- Put fruit and vegetable scraps, peelings and any food that has gone 'off' into a compost bin, Bokashi bin or worm farm, instead of sending it to landfill.

Composting

Composting is nature's way of recycling dead organic matter into new, nutrient-rich soil. Food scraps from your kitchen can be put into a compost heap, along with lawn clippings and other green garden waste. Composting garden waste is covered in more detail in chapter 11, page 203.

Food scraps can also be put into a worm farm, where earthworms break them down and, as in a compost bin, turn them into plant food, this time in the form of nutrient-rich worm castings. Worm farms don't take up much room, so they're a great alternative for an apartment with a balcony but no garden.

Bokashi bins

Bokashi bins are like special food waste processors that use the action of microbes to digest pretty much anything organic in a sealed bin. They can be kept indoors, don't smell and are a great alternative to composting for apartment dwellers and those without a yard.

A Bokashi unit consists of a 20-litre bucket or bin with an airtight lid. The bin has a false floor, which has holes in it to allow drainage and a little air flow, and which sits a few centimetres above the bottom of the bin. There

Getting rid of fats and oils

Unwanted cooking fat and oil should never be poured down the drain or flushed down the toilet. It can cause blockages in plumbing. There are a few options for getting rid of it.

If it's vegetable oil and not contaminated by meat products that might attract vermin, you can wipe small amounts out with old newspaper or paper towel and put that into the compost bin. You can also pour moderate amounts of vegetable oils onto woodchip mulch or sawdust and put that in the compost bin.

For animal fats, pour them into an empty used milk or juice carton, fold down the top and put the lot in with your general rubbish. If it is particularly runny, first put some shredded newspaper, unwanted stale bread or kitty litter into the carton to absorb the bulk of the oil.

If you have a lot of used oil, for example from deep-frying, check with your local council about how they suggest you dispose of it. Some councils or waste management authorities have special collection programs for waste oil and other liquid wastes.

is also a tap at the bottom to drain off any excess liquid, often called 'Bokashi tea'.

Food scraps, including bread and meat, are put in the unit with a little Bokashi mix—bran or sawdust that has been infused with microbes. (Don't worry! They're not pathogens.) You keep adding food waste and Bokashi mix, periodically draining excess liquid, until the bin is full. The waste matter is then allowed an extra week or two to continue fermenting. Then it is buried, adding nutrients to the soil. If you don't have a garden to bury it in, see if you have a friend or family member or local community garden that can take the fermented waste. The waste will initially look quite different from mature compost—more like the food scraps have been pickled—but the waste rapidly breaks down once in the soil. This injection of nutrients is great for plants. The rose bush I planted on top of a buried bucketful of Bokashi-processed food waste is now my most prolific rose bush!

Bokashi units cost around $90, and 1-kilogram bags of Bokashi mix are generally under $10. There are also 120-litre 'wheelie bin' versions designed for commercial kitchens and large workplaces, which cost around $200. Bokashi bins are available from environmental specialty stores and some hardware and nursery retailers.

> ## Compost it!
>
> Here are some of the more unusual items that you can put into your compost bin:
> - vegetable oil
> - tea bags
> - coffee grounds
> - vacuum dust
> - egg shells
> - hair clippings from a haircut, or hair removed from a hairbrush
> - ash from wood fires
> - shredded paper and cardboard
> - dried flower arrangements.

PAPER AND PACKAGING: BEFORE YOU RECYCLE

Packaging has received a lot of bad press in recent years. Packaging makes up much of the mountains of rubbish sent to landfill tips. Much of the rubbish that pollutes our streets and harms wildlife is packaging litter. But the visible drawbacks of packaging are only part of the picture.

Giving credit where credit is due, packaging allows food to be contained and preserved for longer than fresh food would otherwise keep. Packaging means that food can still be eaten when it is out of season, stockpiled in times of plenty and transported to places with inadequate food supplies. But the production of this packaging, like all material goods, has had an environmental cost, as does its disposal. We need to minimise these impacts as much as possible.

Many people focus their efforts on recycling packaging waste after it's been used. However, 'precycling' or avoiding the creation of extra waste by buying wisely in the first place is even better for the environment. You'll often hear this spoken of as the three Rs of waste minimisation:

- **Reduce** your waste-producing behaviour.
- **Reuse** items that would otherwise be rubbish wherever possible.
- **Recycle** materials instead of throwing them away.

Precycling saves you from having to think about recycling or responsible waste disposal later. You can reduce the waste and packaging problem by taking care with what you purchase.

Tips to avoid and reduce waste

- **Buy in bulk**. Buying non-perishable products in bulk quantities means that less packaging is used per unit of product. It's also often cheaper to buy in bulk.
- **Buy the product without the packaging**. Some stores keep goods such as flour, grains and nuts in bulk and allow you to bring your own containers. Similarly, at a bar or restaurant, you can opt for beer or soft drinks in glasses 'from the tap', rather than drinks sold in separate bottles or cans.
- **Buy products that come in refillable or reusable containers**. Many manufacturers now make products and packaging that can be reused. For example, Colgate makes a toothbrush with a replaceable head, and biscuits often come in a retro-style biscuit tin that can be kept and reused.
- **Buy goods in recyclable packaging**. Make sure that the products you buy have recyclable packaging that can be recycled in your local area.
- **Avoid over-packaged products**. Don't buy individually wrapped items or those with unnecessary packaging.

→ ASK TANYA

My wife and I were wondering how to stop unsolicited advertising material being put into our mailbox. We have a very visible no junk mail sticker on the letterbox and every day receive six or more advertising leaflets.

Stephen

I feel your frustration, Stephen! Most of the businesses that distribute advertising mail are happy to abide by people's wishes to avoid junk mail. The last thing they want to do with potential customers is rub them up the wrong way. Unfortunately, there are exceptions.

I've spoken with the Distribution Standards Board (DSB), the self-regulatory arm of the Australian Catalogue Association, which represents the producers of about 90% of all unaddressed advertising catalogues. Their Code of Practice requires that 'no junk mail' stickers be respected.

People can report irresponsible distribution practices by calling the DSB Consumer Hotline on 1800 676 136. The board can follow up the complaint with relevant members. However, a lot of the unwanted junk mail we receive falls into that 10% of materials coming from non-members, who are typically local pizza joints, real estate agents and other local businesses. Contact the business directly and politely inform them that your sticker has been ignored. They will follow this up with their distributors accordingly.

HOUSEHOLD RECYCLING

Recycling takes waste and turns it into a usable material resource that can be made into new products. It's also one of the easiest ways for a person to actively help the environment.

When you put your empty bottles and cans into a recycling bin, you might not think that you're doing very much. You need to remember that thousands of people like you are doing exactly the same thing. In Australia, over three-quarters of our newspapers are recycled.

Making new products from recycled materials uses a lot less energy and water. For example, it takes 75% less energy to make steel from recycled cans than from raw materials, so the overall environmental benefit is huge. However, billions of recyclable cans, jars, bottles, newspapers and boxes are still being sent to landfill when they could be recycled.

There's a bigger picture to recycling than the part we see when we're putting our newspapers, cans, cartons and bottles into our bins. In the last 50 years, the global population has more than doubled, with an estimated 252 people being born every minute. Recycling is no longer optional; it is a necessity if we are to manage our planet's resources sustainably.

Recycling without wasting water

Rinse the food residue out of cans, jars, bottles and drink cartons using old dishwashing water, rather than using fresh water from the tap. Then put them into your recycling bin.

You'd be surprised at the range of items that are made from recycled materials. Many offices now recycle their office paper. Each year literally thousands of tonnes of office paper are recycled into toilet tissue. If each household in Australia substituted four rolls of ordinary toilet tissue with recycled toilet tissue, an extra 10,000 tonnes of office paper could be recycled each year.

In a nutshell, there are two easy ways to bring some environmental action to your home through recycling. First, make sure that you're recycling all that you can in your local area. Second, finish the recycling you started and buy back products made from recycled materials.

What materials can I recycle?

In Australia, local councils provide household recycling services. They or their waste contractors determine the range of materials that are collected, the recycling bins that are used and the frequency of collections. Unfortunately, this means that recycling services can vary greatly from one neighbourhood to another. One council may provide a green waste collection while another doesn't, or one may recycle all plastics, while another might only want plastics numbers 1 and 2.

Remember that just because there's a recycling symbol on a product's packaging, it doesn't necessarily mean that it can be recycled in your area. However, there is a range of materials that are commonly recycled in Australia's major population centres. In regional or remote areas, this range is smaller. The easiest way to make sure that you're recycling all the material that you can is to call your local council to find out about their services, or to visit the relevant page on your council's website. If you live in Australia, visit Planet Ark's online recycling guide at www.recyclingnearyou.com.au.

It's important that you put the right thing in the right bin. A lot of things that shouldn't be put into recycling bins are ending up in them. This makes it harder to sort and recycle the recyclable materials and can also pose a health risk for the garbage collectors. Either through ignorance or irresponsibility, people have placed everything from live grenades and live puppies to used nappies and used syringes in recycling bins. If in doubt, leave it out.

The ultimate recycling guide

On pages 112–13 there is a quick overview of the materials that are commonly recycled through local collections, the types of materials that can be recycled, what they get turned into and some tips for recycling them. It's worth checking with your local council, because not all councils accept the same recyclables.

→ ASK TANYA

I recycle my bottles and jars and other glass. I recently broke an old favourite lead 'crystal' vase. Should I recycle it?

Felicity, Mount Lawley, WA

While I love your enthusiasm, not everything can or should be recycled. (If you're not convinced, go to YouTube and see Celine Dion's recycling of an AC/DC song.) With council-provided recycling collections, limit your glass recycling to just empty bottles and jars. Plate glass, mirrors, drinking glasses, ovenproof glass (Pyrex), vases, crystal and other forms of toughened or heatproof glass all cause problems with the recycling of container (bottle and jar) glass.

Don't waste your waste; reuse it

If a certain material isn't recycled in your local area, it doesn't mean that you can't reuse it. Many rubbish items are being sent to landfill when they could be useful around the house. Here's a list of great ways to reuse some common household rubbish items:

- The cartoon pages of newspapers and old comic books can be reused as a fun gift-wrapping paper for children's presents.
- Stainless steel and chrome household items are currently very popular. Steel food cans with their labels removed are an affordable way to make this look for yourself. Food cans can be used as vases and pencil holders.

Larger food cans and paint tins can be made into umbrella stands and tidy bins. Condensed milk tins with holes punched in them make great tea-light lanterns.

- Small cans, such as baby-food tins, with both ends removed, make great biscuit-cutters.
- Plastic bottles with handles can be made into scoops by cutting the tops off. Two-litre plastic milk bottles in particular make great poo scoopers for those with pet dogs!
- Glass jars are great, see-through storage containers for things like pop-corn kernels, drawing pins, sewing bits and pieces, rubber bands and garden seeds.
- Baby-food jars are good for carrying small portions of salad dressing or milk on picnics. They're also ideal for storing hair clips, hair elastics and safety pins.
- Old electric blankets can be used as under-blankets after the wires have been removed.
- Old towels can be turned into face washers, hand towels or bibs.

PLASTIC BAGS

There's another waste-related by-product of shopping that's become a pet hate. We hate seeing them blowing around the streets, we hate the harm they do to wildlife, and we hate the fact that they'll still be here doing a poor job of biodegrading long after we've gone. Yet somehow, despite this, Australia's 22 million people still manage to use nearly five billion of them every year— tied together, that would be enough to circle the earth over 29 times.

Now Australia's politicians, environmen-talists and retailers are trying to come to some sort of agreement on how we should phase out single-use plastic bags. Australians used an estimated 6 billion plastic bags in 2002, which dropped by 34% to 3.92 billion in 2005. This was partly thanks to the development of the 'green bag'— a strong, convenient reusable bag—and driven perhaps by the threat of a levy. This showed that Australians are more than

Buy right, recycle right

Your grocery choices determine the amount of packaging you buy and therefore determine some of the waste that you will need to dispose of. There are four ways that you can 'buy right' to reduce waste and support recycling:

- Aim for less packaging.
- Make sure that the products you buy have recyclable packaging that can be recycled in your area.
- Buy products that have recycled content in their packaging.
- Buy products made from recycled materials.

Petrol and plastic

Plastic bags are produced from polymers derived from fossil fuels. According to Clean Up Australia, the amount of petroleum used to make a plastic bag would drive a car about 11 metres.

Material	TYPES COMMONLY RECYCLED	RECYCLING ADVICE
Glass	Clear, brown and green bottles and jars	• Don't put ovenproof glass, ceramics, wine and drinking glasses, or light bulbs in your recycling bin as they can cause problems at recycling factories.
Aluminium	Drink cans, foil trays and foil wrap	• 'Foil' potato chip bags should not be included in your recycling bin.
Steel	Steel food cans, pet food cans, aerosols, empty paint tins, coffee tins, bottle tops and jar lids	• Take the plastic lids off aerosols before recycling. • You can see if a can or jar lid is made from recyclable steel by checking with a fridge magnet: if it sticks, it's steel.
Cartons	Milk cartons, juice cartons, aseptic 'brick' cartons	• If your local collection doesn't accept milk cartons, reuse them by giving them to a local school or kindergarten's art room.
Plastic	Types commonly recycled are soft-drink bottle plastic (PET or type 1), milk bottle plastic (HDPE or type 2) and sometimes some opaque PVC bottle plastic (type 3). Some councils accept all rigid plastics, except expanded polystyrene. Note: many supermarkets also collect plastic supermarket bags.	• Leave the lids off plastic bottles • Don't put plastic bags in your recycling bins. Only recycle them through the special recycling bins provided in some supermarkets and retail outlets.
Cardboard	Greeting cards, cereal boxes, larger cardboard boxes	• Many programs don't want cardboard contaminated with food spills, such as pizza boxes. Check with your local council.
Paper	Newspapers, magazines, waste office paper, telephone directories	• Leave out foil gift-wrapping, waxed paper, tissues and self-carbonating paper.

WHAT THEY'RE MADE INTO	DID YOU KNOW ...?
New glass bottles and jars, filtering material, sandblasting material and 'glasphalt' road-fill material	Recycling one glass bottle in the production of new glass containers saves enough energy to power a 100-watt light bulb for four hours.
New aluminium cans	Recycling 20 aluminium cans uses the same amount of energy as it takes to produce one new can from raw materials.
New cans, structural steel for buildings, car parts, bicycles, railway girders and a range of other new steel products	In 2005, Australians had a steel recycling rate of 56%. Every tonne of steel recycled saves 1131 kg of iron ore, 633 kg of coal and 54 kg of limestone.
Office paper, cardboard and fuel briquettes	Milk cartons have lost weight! The amount of material used to make the average milk carton has been reduced over 20% since 1970 through better design.
New soft-drink bottles, fabric, garbage and compost bins, landscaping materials and plastic lumber products	Some kerbside recycling bins are made with up to 50% post-consumer plastic from recycled HDPE bottles.
New cardboard packaging, gift wrap and tissue products	Visy Recycling provided 400,000 pieces of recyclable furniture for the Sydney 2000 Olympic Games, including bookcases and desks made from recycled cardboard.
Cardboard, egg cartons, insulation materials, kitty litter and new newsprint	Making new newsprint from recycled fibres uses six times less energy than making it from virgin pulp. This means that the average Australian household that recycles old newspapers for a year can save enough electricity to power a three-bedroom house for five days.

capable of changing their behaviour when given convenient alternatives and a disincentive (even if only a threat) not to change. But we're starting to go back to our bad bag habits: our plastic bag use for 2007 crept back up to 4.84 billion.

Some supermarkets offer degradable checkout bags. However, these are still a single-use, disposable product and still need resources to be produced. At the checkout, it is better to use reusable green bags than biodegradable single-use plastic bags. In South Australia and, more recently, the Northern Territory, lightweight plastic bags have been banned.

Remember to take any supermarket bags that you do have back to the supermarket for recycling. Most supermarkets in Australia now have special recycling bins for these plastic bags, which then get turned into new plastic products, such as stakes for growing tomatoes.

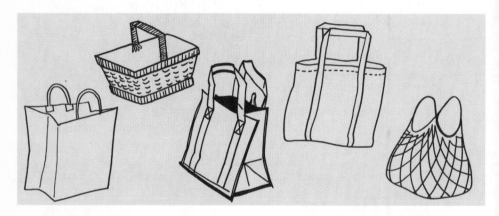

Biodegradable bags

Biodegradable plastic bags made from sustainable materials, such as tapioca or cornstarch, are now available, though not yet widely used. You may see them at the supermarket, offered as an eco-alternative. The *Impacts of Degradable Plastic Bags in Australia* consultancy report to the Australian government investigated the environmental impacts of these plastics and found that 'there is probably little benefit obtained by using biodegradable plastics if you dispose of them to landfill'.

Don't forget your bags!

It's easy to forget to take your reusable bags to the supermarket. Avoid this by putting your bags straight back in the car (if you have one) or by the front door right away after you unpack your shopping.

Strictly speaking, 'biodegradable' means that a substance can be broken down into simpler substances by the activities of living organisms, and therefore is unlikely to persist in the environment. Many people assume that all bioplastics are biodegradable and, conversely, that all petrochemical plastics are not biodegradable. The reality is more complicated. For starters, degradability is not one clearly defined product feature, since plastics can be 'degradable', 'biodegradable', 'photodegradable' and/or 'oxodegradable'. Degradability also depends on where the plastic ends up, be it land, landfill or aquatic environments. Some 'degradable' plastic bags contain little petrochemical plastic bits linked together by substances that break down more easily. In other words, the bag as a whole breaks down into smaller bits of plastic. This plastic dust persists in the environment (microbes in the soil don't know what to do with it) and the bag itself is still made from non-renewable resources. Look for specific statements about what the plastics are made from, their type of degradability and under what conditions. Note that there is an Australian Standard for biodegradable plastics (AS 4736-2006) and a similar European standard (EN 13432) that may be mentioned in statements about biodegradability.

The Plastics and Chemicals Industries Association (PACIA) is working with the Australian government to clear up some of the confusion surrounding plastics and degradability. More detailed information is online at www.pacia.org.au.

Biodegradability can be a cop-out as it allows people to think that their littering or wasteful consumption of plastic bags is okay as long as the plastic is biodegradable. Avoiding disposable bags of any sort by using reusable bags is by far the greener alternative.

LITTER

Litter adds to local pollution, attracts vermin (such as rats) and provides a breeding ground for bacteria. Broken glass and syringes can cause cuts and other puncture wounds. Litter can harm wildlife. Smouldering butts can also be a fire hazard. And it simply doesn't look good.

A little litter goes a long way, given the right conditions. Wind can carry lightweight litter items, such as empty plastic bags, over great distances. Rain can wash the litter off streets into stormwater drains and eventually into rivers, waterways and the seas. In fact it's been estimated that around 95% of the litter on the beaches of Melbourne comes from suburban streets.

Common litter items are cigarette butts, plastic bags, lolly wrappers, bottle tops, straws, cans and bottles, takeaway coffee cups, chip bags and cigarette packaging. Even popped balloons and their strings from balloon releases can end up in the environment as litter and harm wildlife. Litter is also a financial problem, with governments having to spend millions of dollars each year on preventing and cleaning up litter.

Litter tips

- Remember that any dropped rubbish is still litter, even if it's small (bottle lids), it's not noticeable (chewing gum stuck under a seat) or it's biodegradable (an apple core). Use a bin instead.
- Take a small bag or container with you to take litter home in if you're going somewhere where you're unlikely to find a bin.
- You can even go further and pick up other people's litter while you're out and about.

Lethal litter

- Plastic shopping bags are blown into the sea where they do a very good impersonation of a jellyfish. Huge knots of this plastic have been found in the bellies of dead whales and other large marine mammals that have mistaken them for food.
- Fishing line, netting, rope and other litter from sport and commercial fishing can trap and strangle animals.
- The 'honeycomb' plastic that holds a six-pack of cans together is also lethal. The strong plastic rings can strangle and slit the throats of small marine animals, including fairy penguins.
- A post-mortem on a sperm whale found dead on a North American beach found that the whale had starved to death because a plastic gallon bottle had plugged its small intestine.

Cigarette butts

- Cigarettes contain some 3900 chemicals. Many of these are considered dangerous to humans and other living organisms. As filters are designed specifically to trap some of the dangerous by-products of smoking, the cigarette butt is in fact a small poisonous pellet and can kill birds, turtles and other marine animals that swallow them. Studies have also linked cigarette butt pollution to tumours in marine animals. On land, flicked cigarette butts can also cause bushfires, which can kill animals and burn their habitat.
- Butts are made from cellulose acetate (similar to rayon), which is biodegradable in periods ranging from one to two months in favourable aerobic conditions but up to three years or more in seawater. According to the New South Wales Environmental Protection Agency, cigarette butts can take up to 15 years to break down.
- Butts contain not only cigarette tar and nicotine but the residue of complex processes involving hundreds of chemical compounds, many of them bad for your health, including a quantity of radioactive polonium-210.

- Butts littered on footpaths and roads are washed into stormwater drains. An unknown but significant proportion then find their way into waterways and onto beaches. Cigarette butts have been found in the stomachs of birds, sea turtles and other aquatic life.
- Butts may seem small, but their numbers add up. The Cigarette Litter Organisation estimates that 4.5 trillion butts are littered each year worldwide. They are by far the most common litter item found by volunteers on Clean Up Australia Day.
- In Australia it is estimated that the yearly volume of waste from smoking is in excess of 40,000 cubic metres, which you could picture as a 30-kilometre-long queue of some 3000 garbage trucks.

HAZARDOUS WASTE

There's a nasty side to some kinds of waste. Hazardous wastes are those that can pollute or harm the environment or pose a threat to our health. It is important that these wastes are managed by experts in ways that minimise their risks.

A waste is said to be hazardous if it is one or more of the following:
- toxic—harmful or fatal when touched or accidentally eaten. For example, some batteries contain the toxic heavy metal cadmium
- flammable or ignitable—creating fire under certain conditions
- corrosive—containing acids or bases that can corrode metal
- reactive—chemically unstable under 'normal' conditions, reacting easily and able to cause explosions, toxic fumes or vapours when mixed with water.

There are some household waste items that fall into these categories, and there are right ways and wrong ways to deal with them. For solid hazardous wastes, the wrong thing to do is put them in the household recycling bin or rubbish bin and hope for the best. For liquid hazardous wastes, the wrong thing to do is pour them down the sink, as they add to the chemical load that urban sewage treatment plants have to deal with and interfere with the microbial action that is part of sewage treatment. The seriously wrong thing to do is pour harmful liquid wastes into roadside gutters, as these empty directly into creeks, rivers, other waterways or the ocean, with little or no treatment, polluting these aquatic environments.

Household waste items that might be hazardous include the following:
- fluorescent light bulbs (which contain small amounts of mercury)
- solvents, paints and varnish
- used motor oil or car antifreeze
- pesticides

- weedkiller and other garden chemicals
- old mercury-containing thermometers
- old electrical equipment.

If you're throwing out any of these items, there are a number of special collections that will take these wastes and recycle or dispose of them in a responsible way. Look up recycling services in your area at www.recyclingnearyou.com.au, ask your local council, or contact your state or territory waste authority. Some state waste authorities run special collection programs, such as 'Detox Your Home' in Victoria (www.resourcesmart. vic.gov.au) and 'Household Chemical CleanOut' (www.cleanout.com.au) in New South Wales.

MISCELLANEOUS RECYCLING

In addition to the kerbside recycling collections provided by local councils, there are several other recycling programs that collect materials through drop-off points in retail stores, offices, libraries, council buildings and other locations, depending on the material being collected. Electronic waste or 'e-waste', such as computers, mobile phones and printer cartridges, is discussed in detail in chapter 3, page 32.

Planet Ark's 'Recycling Near You' website (www.recyclingnearyou.com. au) allows people to type in their postcode or council area and find details of recycling services available in their area. The site includes information on collection and recycling programs for corks, car batteries, car tyres, gas cylinders, light globes, motor oil, medicines and scrap metals.

Your home refuse may also include household goods that are in good working order or reasonable condition and that could be reused by someone else. This may be children's clothes that are outgrown before they wear out, your unwanted clothes, bric-a-brac and so on. If they're in good condition, they can be donated to a charity that will redistribute them to people in need or sell them and use the money raised to do good work. However, charities shouldn't be used as a garbage-disposal service. It costs them money to get rid of items that are broken or in too poor a condition to sell, diverting funds from the important work they do.

Another way to give away household goods, or to receive some yourself, is through the Freecycle Network, a web-based network in which members post either items they want or items they're happy to give away. Visit www. freecycle.org and type in your city, state and country.

Or if you want to turn unwanted goods into cash, there's always the good old-fashioned garage sale, newspaper classifieds, and online second-hand trade sites like eBay.

→ ASK TANYA

What are the options for getting rid of old mattresses? We want to replace our ten-year-old inner-spring but don't know what to do with it.

Angela, Manly, Queensland

Option 1: Take the old mattress to the tip. But with the average mattress weighing 35 kilograms and taking up 0.7 cubic metres of landfill space, landfill operators (who aim to compact waste to about 1 tonne per cubic metre) would say they're literally a waste of space.

Option 2: Donate the old mattress to a charity, complete with dust particles, dead skins cells, dust mites, bacteria and mould spores—eek!

Option 3: Recycle it. Dreamsafe Recycling accepts old mattresses for recycling in Adelaide, Canberra, Geelong, Melbourne, Sydney and Wollongong, with plans to open facilities in other major cities. Call them on 1300 551 245 or visit www.dreamsafe.com.au—while you're online, have a look at the highly amusing 'Classic Mattresses' photographs. They can take mattresses directly from the public (for a fee of around $25) but also operate through council and retail collections. Dreamsafe Recycling cleans, sanitises and deodorises those mattresses in good enough condition and distributes them to charities in a far more hygienic state. Other mattresses are recycled, recovering wood, foam, wadding, springs and some fabrics.

If you're buying a new mattress, ask the retailer if they will take back your old one and recycle it. If they won't and Dreamsafe don't collect in your area, contact your local council or state government waste authority and ask how they would prefer you to dispose of the mattress. They may tell you to take it to a tip or waste transfer station (usually with a fee of $20–25). Then it will either go to landfill or get passed on to a recycler.

MAKE A COMMITMENT

This week I will...

☐ _____

☐ _____

☐ _____

This month I will...

☐ _____

☐ _____

☐ _____

When I get the opportunity I will...

☐ _____

☐ _____

☐ _____

Notes

_____

_____

_____

_____

_____

_____

9

Energy

Energy is defined as the ability to do work. And what wonderful work energy does for us. It boils the water to make your morning cuppa, airconditions sweltering offices, provides light after dark and powers the stereo. All of nature interacts with the free energy from the sun, and using the warmth of passive solar energy is part of being human.

Energy—in the forms of electricity, gas, LPG and other fuels—is an integral part of our homes. We use it to do a range of useful and important things, like cooking food or refrigerating it. But as well as the practical home uses of energy, there are also the fun uses. Electricity powers televisions, computers, fairy lights, game consoles, cordless telephones, lava lamps, hair-straightening irons and many other devices. Even the humble open fire may be more about romance than home heating.

Our environmental concerns relating to energy boil down to two issues: the energy we currently use has serious impacts, such as climate change, and we also need to ensure our future 'energy security'—that we have enough sources of energy to keep enjoying its benefits and uses into the future.

In the first part of this chapter, we look at energy use in our homes and how we can cut it back and still live comfortable, enjoyable lives. In the second part, we look at where this energy comes from and how we can source the energy we do use from cleaner, renewable sources.

→ **ASK TANYA**

I currently buy 100% green electricity. I have recently noticed lots of advertising promoting reducing energy usage at home and at work. Is there any reason, other than reducing my bill, for me to cut down on my usage?

Peter, Orange, NSW

First, a hearty 'good on you' for being a GreenPower subscriber. But your chosen source of electricity isn't the only way you influence Australia's electricity situation. Your electricity usage, along with that of all the other grid-connected electricity users, places a demand on power stations. New electricity-generation projects are being built to keep up with our increasing collective energy demand.

Power stations, renewable energy projects, powerlines and other energy infrastructure all have environmental and financial costs (some much more than others). Queensland Premier Anna Bligh summed it up when in June 2009 she announced the state's ban on the sale of inefficient airconditioners: 'Every time an air-conditioning unit is installed, it costs our network up to $5000'.

Remember also that greener electricity sources are not completely without environmental impacts. For example, hydro projects can affect aquatic ecosystems, and wind farms can harm local wildlife (though I'd hazard a guess that more birds have been killed by cars, tall buildings and pet cats than wind turbines). In short, unlimited consumption is never sustainable—even an all-you-can-eat restaurant has a door price, and poses a risk of overeating, indigestion and weight gain!

ENERGY IN THE HOME

There are four steps in a simple household energy action plan that aims to cut the carbon impact of your energy use:

1 **Measure** your energy use. This is easy. Your electricity and gas bills do this for you. Monitoring your usage will help you to keep track of how effective your efforts to reduce your energy use are. Keep in mind that electricity prices are increasing, so look for changes in average daily energy use (measured in units, kilowatt hours or megajoules) rather than changes in the dollar figure to get a true indication of how you're going environmentally.

2 **Reduce** your energy use. That's what this chapter is all about.

3 For the energy you do use, go **renewable**, as discussed in the second part of this chapter.

4 **Offset** the carbon impact of any non-renewable but less greenhouse-intensive energy you do use, such as natural gas used for home heating. This also is covered later in the chapter.

Where does our home energy come from?

Most of the energy we use in our homes comes directly from burning fuel such as firewood or natural gas, or indirectly in the form of electricity. The natural gas and LPG used in homes are both fossil fuels. In Australia over 90% of electricity is generated by burning coal and natural gas—both fossil fuels—at power stations. In this chapter, I refer to this as 'conventional electricity', as it's often the default electricity source. It's generally what you get unless you ask to switch over to GreenPower or unless you generate your own—for example, with solar panels. By comparison, more than half of New Zealand's electricity is generated by hydroelectric schemes. The average Australian household contributes more than 15 tonnes of greenhouse gases each year by using electricity and natural gas, driving cars and producing waste. If you add the emissions that result from producing the goods we consume, the figure is much higher. Some of this greenhouse gas is produced in the home—for example, the carbon dioxide from an open fire.

However, much of it is produced away from the home, at the power stations that provide electricity or at the factories that produce the products we buy. In this chapter, we will be focusing on the greenhouse emissions from our direct use of energy in our homes.

Different energy sources make widely differing contributions to climate change, so it's useful to talk about the idea of 'greenhouse intensity'. This is a measure of the emissions that result from producing a given amount of energy. Our electricity in Australia is at the high end of the greenhouse intensity scale when compared with other countries, largely due to our reliance on coal. As a rough guide, a kilowatt hour of electricity produced in Australia means a kilogram of carbon dioxide released into the atmosphere. Note that this is a national average. In Tasmania, the emissions intensity of electricity is about 0.32 of a kilogram of carbon dioxide per kilowatt hour because of the use of hydroelectricity, while in Victoria the use of polluting brown coal results in an average electricity emission intensity of 1.23 kilograms of carbon dioxide per kilowatt hour. The higher the proportion of renewable electricity fed into our national grid, the lower the greenhouse emissions of electricity use will be. In this book, emissions figures quoted for electricity use are based on the national average (which largely comes from polluting fossil fuels), unless stated otherwise.

Similarly, we can talk about the greenhouse intensity of using natural gas. Because it is burned directly in our homes for cooking or heating, it produces less emissions than would be produced by the equivalent amount of conventional electricity needed to do the same job. For cooking, heating and providing hot water, you have a choice between gas and electricity, provided you have access to natural gas. You also have a choice in relation to electricity source. Keep this is mind as you work through this chapter. As a rule of thumb, natural gas produces less greenhouse emissions than conventional electricity, and renewable electricity produces less emissions again.

Natural gas is not available everywhere, but areas without natural-gas infrastructure can use LPG instead. As fuels, they are similar environmentally. However, LPG needs to be stored under pressure and must be transported by tanker or in cylinders, adding to its environmental impact. LPG is also around double the cost of natural gas.

Where do we use energy?
The place to start saving energy is by developing an understanding of where we use it in our homes. The following graph shows the main energy-using activities and the greenhouse emissions that result from them. The fact that the two columns don't duplicate each other demonstrates the difference

that various energy options can make. Virtually all appliances, except gas cookers, run on electricity, which is why appliance use has disproportionately high emissions. Conversely, space heating and cooling represent 38% of our energy use but only 20% of our household emissions, largely because a lot of our heating runs on natural gas. This is an important consideration to keep in mind when choosing products such as cookers or heaters, which come in both gas- and electricity-run models. Producing heat with natural gas is generally more efficient than heating with electricity using a resistance element.

Energy use and greenhouse emissions of the average Australian household

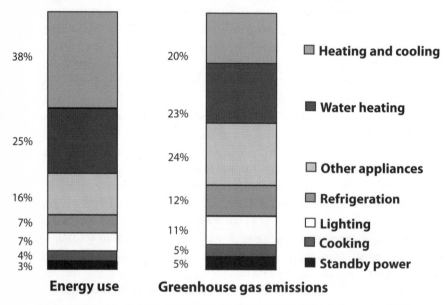

Energy use   Greenhouse gas emissions

Source: data from Commonwealth of Australia, *Your Home Technical Manual*, 4th edn, Department of Climate Change and Energy Efficiency, Canberra, 2010

The graph shows the areas that are the 'big hitters' when it comes to energy use: heating and cooling, water heating and appliances. This chapter will address them in that order. Think about the 'big hitters' as you set your priorities, and target first the areas where you'll get the greatest result from your efforts. For example, insulating your home to reduce your energy needs for heating and cooling can potentially save a lot more energy than putting lids on saucepans to reduce the energy needed to cook (though the little things do add up).

ENERGY RATING PROGRAMS

To make sense of the advice in the coming pages, you'll first need to understand star ratings and Energy Rating labels. Modern homes in Australia have a plethora of energy-using appliances and equipment, from ducted heaters for the whole house to single light bulbs, with hundreds of different types of appliances and gadgets in between. Some use more energy to run than others. Generally, appearances won't tell you how efficient a piece of equipment is. Fortunately, there are energy-efficiency labels, backed by government regulation, for certain types of products that can help you to identify energy-wise models.

The Energy Rating label

If you've been shopping for white-goods lately, you may find that you're seeing stars—at least stickers with stars on them. This is the Energy Rating label. The appliance Energy Rating label scheme is a joint initiative of Commonwealth, state and territory government agencies and the Energy Efficiency and Conservation Authority of New Zealand. It gives the public information about how much energy particular appliances consume. This initiative helps people who are shopping for appliances to make an informed choice. Appliances are rated for energy efficiency on a scale of one to six stars.

What does a kilogram of carbon dioxide look like?

Many people find the greenhouse effect hard to understand. Part of the reason for this is that greenhouse gases are mostly invisible. Those that come from our electricity use are often released at power stations far away from our homes. The challenge for environmentalists is to find a way for ordinary people to visualise the greenhouse effect.

You will often find quantities of greenhouse gases expressed in terms of carbon dioxide equivalents, calculated by multiplying the emissions of each gas by its relative contribution to global warming. This is measured as a weight. State government agency Sustainability Victoria has come up with a way to help people visualise and understand different amounts of carbon dioxide. Staff at Sustainability Victoria worked out that it takes about 50 grams of carbon dioxide to fill a party balloon. They have now adopted 'the black balloon' as a unit of measurement for greenhouse gases:

1 black balloon = 50 grams of carbon dioxide (equivalent)

For example:

- Washing in cold water instead of hot saves the average household 2531 black balloons of carbon dioxide per year.
- Drying one load of clothes with an electric clothes dryer = over 60 black balloons (or over 3 kilograms of carbon dioxide).
- Drying clothes on a clothes line = none.

The more stars a product has, the more efficient it is, the less electricity it needs to do its job and the lower its ongoing running costs.

The rating scheme covers dishwashers, washing machines, refrigerators, freezers, clothes dryers, airconditioners and, more recently, televisions. The

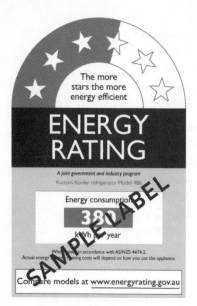

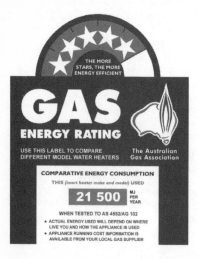

program is mandatory, meaning that all of these types of appliances are required by law to have an Energy Rating label.

The Gas Industry runs a similar program that rates the energy efficiency of appliances that run on gas. The Gas Energy Rating label covers gas space heaters, gas water heaters, gas cookers and outdoor gas barbecues.

Choose energy-efficient appliances with five or six stars. You'll save money and energy every time you use them.

For more information about the Energy Rating label, go to www.energyrating.gov.au.

The second price tag The Energy Rating label also tells you how much energy an electric appliance uses each year in kilowatt hours (kWh) when tested to the Australian Standard. The smaller this number is, the better the appliance is for both the environment and your energy bills. You can use this number to estimate the running costs of a given appliance. For example, a 4.5-star fridge that uses 506 kilowatt hours per year running on electricity costing $0.18 per kilowatt hour will cost around $91 in electricity to run for a year. Over a life span of 15 years, that's over $1365. By comparison, a 3.5-star fridge using 738 kilowatt hours per year will cost nearly $133 per year to run and over $1990 over 15 years. In this case, an Energy Rating of only a single extra star will save you over $620 on running costs over a 15-year life span.

The running cost of an appliance over its lifetime is often referred to as the 'second price tag', with the initial outlay of the purchase price termed the 'first price tag'. The Gas Energy Rating label similarly shows how much gas an appliance typically uses in megajoules (MJ) per year. Calculate and consider the second price tag when you compare appliances, and remember that when it comes to energy-efficient products, saving money goes along with saving the environment.

The Energy Rating initiative has an online calculator that allows you to compare the energy ratings for various products and easily calculate the second price tag. Visit search.energyrating.gov.au.

KEEPING WARM

The first thing many of us spend money on when we move into a new house is homewares and soft furnishings such as plush throws and cute cushions. But what's the use of looking fabulous if you're freezing? The benefit of a greener, more energy-efficient home is that it's more comfortable. So get comfortable first, and decorate later.

Many of us unwittingly spend heaps of money on electricity and gas to heat and cool our homes, only to have the heat escape out the window. Some people pump even more money into replacing perfectly good heating systems, and still find that their homes don't stay a comfortable temperature. You can spend thousands on a new heating and cooling system, and hundreds more on energy to run it, but you'll be throwing your money away if your home isn't airtight and well insulated.

Heating and cooling account for around 38% of the average Australian home's energy use (electricity and natural gas). This can rise to 60% in cooler areas where more heating is needed. Consequently, this energy use accounts for a good part of our greenhouse gas emissions. This can be greatly reduced by stopping winter warmth from escaping, and summer heat from coming in. In fact, in many parts of Australia, preventative measures and good house design alone can keep an entire house comfortable for most of the year, with little need for additional heating or cooling.

Your heating and cooling costs and your impact on the environment can be greatly reduced if your home is well insulated. Insulation helps to reduce the unwanted movement of heat into and out of your home. Heat can escape through uninsulated ceilings, walls, floors and uncovered windows. Draughts and other air leaks can also allow heat to escape and cold air to come in.

The diagram of winter heat loss on page 128 shows you how heat can escape. You can address each area and make simple changes to ensure that your home stays warm. Many of these changes are easy to make and will reduce the need for additional heat.

Controlling heat loss

- Add weather-stripping to exterior doors to seal the gap between the door and its frame. You can also use a door sausage at the base of the door.

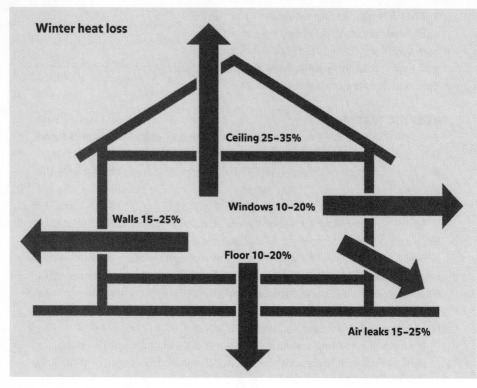

Winter heat loss

Ceiling 25–35%

Windows 10–20%

Walls 15–25%

Floor 10–20%

Air leaks 15–25%

Also add weather-stripping to attic doors or roof space hatches, particularly if the attic or roof space is poorly insulated.

- Locate any air leaks and draughts using a lit candle or the flame of a cigarette lighter. Slowly move it along skirting boards and other joins; draughts will make the flame flicker. Alternatively, use a downy feather or thin piece of tissue if you're concerned about the fire risk of using a flame. Once you've found the gaps, caulk or seal them with a filler product. Ask your hardware store for advice on the right product for the job.

- Do an outdoor inspection as well as an indoor inspection for gaps and cracks. Carefully look over the house's exterior surfaces, particularly around window and door frames.

- Install dampers in chimneys. Block off the chimneys of any fireplaces that you don't use.

Filling in the gaps

- **Caulking** is filling a gap between two materials that meet but do not move relative to one another.
- **Weather-stripping** is closing a gap between two materials where one surface moves in relation to the other—for example, the gap between a door frame and a door.
- **Sealing** is making an airtight (sometimes watertight) barrier or closure between two spaces—for example, around plumbing holes and pipes.

- Install self-closing exhaust fans.
- Choose windows with low-emissivity (low-e) coatings, which help to reduce the heat loss through window panes.
- If you are using a room heater, keep the doors to other unused rooms closed, especially those with exhaust fans that open to the outside, such as the bathroom.
- Keep the curtains of north-facing windows open during the day to capture heat from the sun. Once the sun has gone down, close the curtains to keep heat in.
- Choose window coverings that will help to prevent draughts and keep the heat in.
- Multiple layers of glazing improve a window's 'thermal resistance' (its ability to block the movement of heat). Consider fitting an additional pane of glass or clear acrylic to existing windows. You can get DIY double-glazing retrofit kits from hardware stores and environmental retailers. If windows need replacing, install double-glazed windows or glass bricks, which have a similar heat-efficiency effect to double-glazing.
- Assess your home's insulation and consider increasing the level of insulation if it's currently inadequate. Ceilings in Australian homes should be insulated to at least R2.5 level (see page 235) and, if possible, the walls as well. If you're building in a cold climate, you may even wish to install insulation under the floor.
- If your walls are made from empty framework without insulation, then consider filling the cavity with blow-in insulation.
- Rubber provides the best seal for weather-stripping and is waterproof. Rubber may be more expensive, but it lasts longer, is more effective and easier to clean than foam, felt and brush edge weather-stripping.
- Milk or mail chutes in the front door, laundry chutes (particularly those that end in an uninsulated basement utility room), firewood chutes and pet doors can all let a huge amount of heat escape. Close them up completely if you don't use them.
- Check for gaps or cracks in external walls, interior walls, floors, ceilings and joinery at the end of winter and again at the end of summer. The changes in temperature and moisture levels can cause some building materials to expand or contract, causing new gaps to form. Your hardware store will have caulking products to seal these gaps.
- If you have a fixed evaporative cooler installed, check if there are draughts coming through the vents in winter. You can reduce these draughts by fitting an external cover over the ceiling vents, which can be removed in summer.

Energy-efficient window dressing

Some of your winter heating, and the money you spend on it, could literally be going out the window. Around 10–20% of the heat loss of the average home is through windows. This heat loss can be as high as 30% in poorly designed homes with large windows.

You can reduce the loss of heat through your windows, and the gain of heat in summer, by choosing your window coverings carefully.

- Choose drapes made from tightly woven fabrics.
- Line curtains with a reflective backing to reflect summer sun.
- Have curtains made to a snug fit for the windows. Make sure they are wide enough to stretch to both sides of the window.
- Choose curtain rails or tracks that curve around at each end, bringing the edges of the curtain right up to the wall.
- Place boxed pelmets over the top of the curtain.
- Make sure that the curtains are long enough to fall well below the bottom of the window. Consider floor-length curtains where possible.
- If you choose holland or roman blinds, make sure that they are made from thick, closely woven fabrics and that they are well fitted to the window.

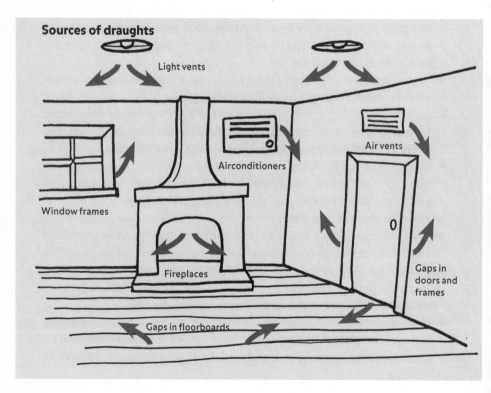

Sources of draughts

Light vents

Airconditioners

Air vents

Window frames

Fireplaces

Gaps in doors and frames

Gaps in floorboards

- Louvre, venetian, timber venetian and vertical blinds are not good insulators. Non-metal blinds can effectively block summer heat, but do not significantly prevent heat loss in winter. They are best used in warm climates or in the homes of severe asthmatics, where fabric curtains can increase dust and dust mite levels. Metal blinds on the sunny side of the house can conduct heat into the room in summer.

THE ULTIMATE HEATING GUIDE

The ways in which you can retain heat within your house and better use your existing heating system are called passive measures. Not everyone has a well-designed, energy-efficient house that doesn't need additional heating. Nor do we all have the immediate opportunity to build one. This is when we look at heating systems to help us keep warm in winter.

First, take a good look at your house, the people who live in it and how they use it. This will affect your heating needs. For example, a family of four building a new home may decide to get in-slab heating or ducted heating to a number of rooms, while a single person renting a flat may only have the option of portable room heaters. Make sure you've exhausted the options of insulating better and blocking draughts, then ask yourself if you really need heating or whether passive measures will suffice. Then decide which rooms in your house need frequent heating; consider how big they are, how often they need to be heated and for how long. If possible, make sure heated rooms or zones of the house can be closed off from the rest of the house and are well insulated. The next step is to decide which heating system best suits your needs, is better for the environment and is more economical.

Tips for more efficient heating
- Wear a jumper! Central heating wasn't invented so that you could set the thermostat to 25°C (77°F) in winter and wear a T-shirt. Instead of turning on or increasing the heating, wear clothes that are appropriate for the season.

- Set the thermostat to a reasonable temperature of 18–20°C (64–68°F). For every degree (Celsius) you increase the thermostat setting, your energy bill will increase by up to 15%.
- Regularly clean any heating ducts or filters. This will allow them to operate more efficiently, reduce the build-up of dust and other allergy triggers, and help to reduce fire risk.
- If you're finding heat is accumulating around the ceiling, leaving the lower part of the room cold, install a ceiling fan to circulate warm air. A ceiling fan with a reverse function can help convective heaters to heat more efficiently. When the fan turns slowly in the reverse direction, a gentle updraft is created. This re-circulates the hot air trapped at the ceiling to provide even heat throughout the room.
- If you live in a two-storey house, use any upstairs living areas in winter, rather than heating downstairs living rooms. Upper storeys will be warmer because hot air rises.
- Heat only the rooms that people are in. Close any heating ducts in unused rooms. Use a small portable room heater if only one room is being used.
- Close the windows and doors of rooms where a heater is on.

Types of heating systems

There are two basic types of heating: radiant and convective. They can be used in different ways to heat a single room, a zone of your house or the entire house.

Radiant heaters heat people and objects by the direct radiation of heat from a hot surface. Examples include bar radiators, heated concrete slabs and open fires.

Convective heaters transfer heat by warming air and circulating it. Personal fan heaters, reverse cycle heat pumps, electric panel convectors and ducted central heating work this way.

Space or **room heaters** warm a single room or zone of a house, rather than the whole house.

Central-heating systems produce heat at a central point and distribute it as heated air through ducts or as heated water or oil through pipes, potentially to the whole house.

In-slab heating is where special electric cables or hot water pipes (**hydronic floor heating**) are laid into the house's concrete slab. This system needs to be incorporated into the slab as the concrete is poured, so it is usually only considered for new homes or extensions. Heat is either produced directly in the slab in electric cables or in a central hot water system, which then pumps the heated water through pipes laid in the slab. Either way, the slab produces radiant heat.

Hydronic heating heats water in a boiler and then pipes it around the home to radiator panels, skirting board convectors or fan coil convectors. These heat the destination room through convection and radiant heat. Systems typically have a central boiler, a series of pipes and radiator panels. If using a gas-fired system, a low water-content boiler will be more efficient. Storage boilers have slightly higher running costs due to continual 'standing' heat loss from their tanks. Many hydronic systems can also be used to provide hot water for other uses in the house, though most perform this task less efficiently than dedicated hot water systems. These are often referred to as 'combination' boilers. In addition, they can be linked to swimming pool heating systems. Solar hot water systems can also be used to provide the hot water, greatly reducing the running costs and environmental impacts. Hydronic heating has been used for nearly a century in Northern Europe and North America.

Reverse-cycle airconditioners or **heat pumps** use electricity to move heat from a source (the air or the ground) into your house, in the same way a fridge or airconditioner uses electricity to move heat out. It takes less electricity to heat a room this way than to use electricity to directly generate the heat within the home. Heat pumps operate at optimum efficiency in areas with mild temperatures. They are less efficient when the temperature of the air or ground source area drops below 4°C or rises above 38°C. As the name suggests, **air-source heat pumps** take heat from the air outside the house and are the more common type of reverse-cycle airconditioner used in Australia. **Ground-source heat pumps** (also known as 'geo-exchange heat pumps') draw heat from the earth or groundwater or both. As such, they are often called 'earth-energy' systems. Underground temperatures are mild and more constant, making ground-source heat pumps more efficient and more effective, without the performance drop in extreme weather. However, the ground needs to be excavated to install ground-source systems. Consequently, ground-source systems aren't practical for existing homes but are an option for new homes, be it a somewhat expensive one.

The reverse function of a heat pump provides refrigerated air. In fact, airconditioners were the first form of this technology to be commonly used. Although heat pumps can reduce heating costs, many people who buy them find that their electricity bills actually increase over time. This is because many people who purchase a heat pump with a cooling function did not previously have an airconditioner. The increase in electricity use is due to the additional power used for cooling in warmer months. If you have an airconditioner, resist the temptation to use it other than in exceptionally hot weather.

General rules for heating systems

While there are a range of different heaters and heating systems, with individual models having their own features and efficiencies, a few generalisations can be made. The size and shape of a room will determine the type of heater that will most efficiently heat it. Because convective heaters warm the air itself, rooms with a lot of air (in other words, large rooms) or with open stairwells and other openings are not suited to convective heaters. Remember also that hot air rises, so a convective heater in a room with a high ceiling will not provide heat efficiently for people at ground level, although a ceiling fan can help to recirculate the warm air that collects at ceiling level.

- Convective (air) heaters are better for heating small rooms quickly. They are not very effective at heating large spaces.
- Radiant heaters can take a while to warm up, but are better for larger spaces and draughty rooms.
- Pilot lights in gas heating systems can cost up to $25 per year extra. Choose systems with electrical ignition instead, or at least turn the pilot light off in summer when the heater is not in use.
- Unflued gas heaters release unburnt gas and combustion gases into the home, contributing to indoor air pollution. They also tend to cause a lot of condensation.
- Only use radiant heaters in bathrooms and ensuites, as the moving air of convective heaters has a cooling effect on wet skin. You can now get bathroom light fixtures that incorporate an infrared heat lamp and exhaust fan.
- Gas heating and efficient reverse-cycle electric heat pumps are more efficient, accounting for a third of the amount of greenhouse emissions of other electric heating systems that use resistance elements.

Comparing central heating systems

It's worth considering the main types of central heating systems, taking into account the purchase and installation costs, the ongoing running costs, the greenhouse gas emissions, any safety issues and where they are best installed. The running costs and emissions given on the following pages are based on the more efficient models of a given system and are intended to indicate the relative cost of each option. Also remember that the greenhouse emissions for electric heaters will be lower than stated if you're using green electricity through GreenPower programs.

	DUCTED GAS CENTRAL HEATING	REVERSE-CYCLE ELECTRIC HEAT PUMPS	ELECTRIC IN-SLAB FLOOR HEATING	HYDRONIC HEATING
Up-front costs	From $2500 for a basic 6-outlet system plus installation costs	$6000–$15000 (includes installation)	$35–45 per square metre installed	From $5500
Running costs	Low–medium, depending on the efficiency of the heater	Medium	High	None (solar-heated water) Low (gas-heated water)
Green-house emissions	Low Up to 2.1 tonnes CO_2/year	Medium 6 tonnes CO_2/year	High 14.7 tonnes CO_2/year	None (solar-heated water) Low (gas-heated water)—up to 2.5 tonnes CO_2/year
Safety	Childproof	Childproof—fuse-protected installations	Childproof	Childproof
Notes	• This system is more effective if vents are in the floor, but floor vents generally can't be fitted after the house has been built	• Heats the room quickly • Can provide cooling in summer	• Slab heating is not suited to providing occasional short bursts of heat, such as half an hour on a high setting to take the chilly edge off • Suitable for people suffering allergies or respiratory complaints	• Systems that provide hot water as well as heating will add approximately $400–500 to the cost of a low-water content boiler and up to $900 to the cost of a storage boiler
Look for	• Systems that can be zoned to heat different areas separately • Well-insulated ducts (Australian Standards R1.0 to R1.5) • Electronic ignition rather than a pilot light • Programmable timer and thermostat controls • A high star rating (four–five stars on the Energy Rating label)	• Energy-saving inverter technology • Systems that can be zoned to heat different areas separately • Well-insulated ducts (Australian Standards R1.0 to R1.5) • Programmable timer and thermostat controls. • A high star rating (four–six stars on the Energy Rating label). Note that there are two ratings for reverse-cycle systems, showing the efficiency in both heating and cooling mode	• Adequate insulation for the slab, including the slab edges • Individual thermostat control for each room and/or zone controls • Programmable timer	• Independent valve controls/thermostats in each room • Systems with solar thermal collection panels to pre-heat the water

Comparing fixed space heaters

Space heaters can be used to heat one or two rooms. They are cheaper than installing a ducted system for the whole house and can easily be fitted into an existing house. Some space heaters are fixed; others can be moved from room to room. Portable models may be your only option if you're renting and have a disagreeable landlord.

	REVERSE-CYCLE AIRCONDITIONERS	SLOW COMBUSTION WOOD HEATERS
Up-front costs	• Portable—$800–3000 • Fixed (window/wall)—$500–1600, plus installation • Split—$1000–10,000, plus installation	From $800, plus installation
Running costs	Low–medium	Low
Greenhouse emissions	Low–medium 2.7 tonnes CO_2/year	Low 1.3 tonnes CO_2/year
Safety	Childproof	• Surfaces of wood heaters become very hot, posing a burn risk • Fire risk from sparks and embers when stoking the fire or adding wood
Notes	• All types of airconditioners have a compressor, which is usually outside the house. Compressors are generally noisy. • These come in single wall-mounted, split and multi-split systems. • Multi-split systems have 2–7 interior diffusers running off 1 exterior unit. In such systems, many rooms can be heated.	• They can heat up to 2 rooms.
Look for	• Energy-saving inverter technology • Remote thermostat and programmable timer • Adjustable directional louvres • A high star rating (4–6 stars on the Energy Rating label)—note that there are 2 ratings for reverse-cycle systems, showing the efficiency in both heating and cooling mode	• Wood fuel from sustainable sources only • Air intake controls

OPEN FIREPLACES AND WOOD HEATERS	CONVENTIONAL ELECTRIC HEATERS	GAS WALL HEATERS, STOVES AND 'LOG' FIREPLACE INSERTS
From $1800, plus installation, assuming you already have a chimney	$200–500	From $1000, plus installation
Medium	High	Low
High, plus combustion emissions	High 6 tonnes CO_2/year	Low 1.1 tonnes CO_2/year
• Embers and sparks are a fire risk—a screen for open fires is essential. • Surfaces of wood heaters become very hot, posing a burn risk. • Combustion gases and particles may contribute to poor air quality and pose a health risk to people with respiratory sensitivities.	• Surfaces become warm, but not too hot.	• Surfaces of wall units become warm, but not too hot. • Exposed flames can be a fire and safety risk, particularly with small children. • Flues should be periodically checked—faulty flues can cause carbon monoxide poisoning.
• These are romantic and look gorgeous, but are very inefficient—up to 90% of the heat goes up the chimney. • They can heat only 1 room.	• Wall panels use peak electricity and are expensive to run. • They are not very effective at heating large areas.	• They can heat up to 2 rooms. • Gas log fires provide the atmosphere of an open fire without the ash, solid fuel or high greenhouse emissions.
• A chimney with a damper to prevent air leaks when the fireplace isn't in use (important) • Wood fuel from sustainable sources only	• Wall panel convectors and off-peak heat storage systems • Remote thermostat and programmable timer	• A high star rating (5–6 stars) • Child locks on closed systems • Remote thermostat and programmable timer • Heat outlet at floor level • Oxygen depletion and overheating sensors (desirable) • 'Balanced' or 'power' flues • Electronic ignition rather than a pilot light

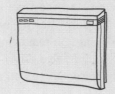

Comparing portable heaters

	ELECTRIC FAN HEATERS	ELECTRIC CONVECTION HEATERS
Up-front costs	From $40	From $70
Running Costs	High	High
Greenhouse emissions	High	High
Safety	• These are reasonably safe—keep away from water to avoid electric shock. • Those with a ceramic element are generally safer as the element doesn't get hot enough to start a fire.	• Be careful with hot surfaces.
Notes	• They produce heat very quickly but are not very effective at heating a large open room. • They are cheap to buy but expensive to run.	• They are not suited to larger rooms with high ceilings or poor insulation. • They can have fans as well.
Look for	• Automatic shut-off functions that switch the heater off if it's tipped over • Units with thermostats	• Units with thermostats

Portable space heaters can be moved from room to room as they're needed. They are a good option if you're renting and you can't convince a stingy landlord to install a better heating system. Once you've bought one, you can take it with you if you move house.

Heating with wood

There's nothing quite like a glowing fireplace to provide a warm, relaxing and romantic atmosphere. However, there are good and bad wood heaters. Older-style wood heaters and the traditional brick fireplace are very ineffective. Some even rob heat by drawing heated air up and out the chimney while inadequately heating the remaining indoor air. They burn through a lot of wood for only a little heat. They can also produce harmful combustion gases, including poisonous carbon monoxide, and emit fine

RADIATORS	OIL COLUMN HEATERS	UNFLUED OR 'VENT-FREE' GAS OR KEROSENE HEATERS
From $40	$80–400	$400–1500
High	High	Low–Medium
High	High	Low
• Hot surfaces can cause burns. • These can also be a fire risk, particularly if near draped, flammable clothes or curtains. • Radiators that use a quartz tube pose a fire hazard as the quartz tube can be easily broken.		• These pose a risk to indoor air quality through keeping emissions and moisture in the room and encouraging mould growth.
• They're good at spot heating (focusing heat on a small area). • They are cheap to buy but expensive to run. • Fixed radiators are good for bathrooms as they directly heat your body without the cooling effect of moving air on damp skin.	• If you have to choose a portable option, oil column heaters are better suited to larger rooms with high ceilings than radiators or electric fan heaters. • They are slow to respond but hold heat well. • They can have fans as well.	• The retention of harmful combustion by-products, including carbon monoxide, is a health risk, particularly in winter when people tend to air their houses less frequently.
• Safety features and designs that prevent tipping over 	• Thermostats and timers	• Look for a different type of heater!

combustion particles that can penetrate deep into the lungs, causing respiratory irritations. Those particles and emissions that escape into your home are unhealthy for you and your housemates, while those vented through the chimney pollute your neighbourhood. Also of environmental concern is the wood source, which may be old-growth forest.

The good news is that there are good wood heaters available too. Modern slow combustion woodstoves, fireplaces and fireplace inserts allow you to heat with wood with greater efficiency, lower fuel costs and less pollution. When you do use an open fire or a wood heater, there are a number of things that you can do to reduce the amount of polluting smoke produced. Only use good-quality, dry firewood or fuel briquettes. Wood needs to be clean and to have been dried (or seasoned). Wet or green wood is harder to burn and produces more polluting gases. Never use wood that has been treated with chemical finishes or varnishes or pressure treated. Never burn particleboard, plywood or other wood composites, which contain glues and other bonding agents. Burning these woods can produce toxic gases.

You should also remember the following when heating with wood:

- Use wood that comes from sustainable sources. Don't take dead fallen logs from parks or forests, as they provide homes for wildlife. Don't use driftwood because it may contain chemical treatments or preservatives.
- Don't burn glossy magazines, plastic, cardboard or household garbage.
- Consider making your own fuel briquettes. You can buy fuel brick-making devices from hardware stores that enable you to make briquettes from old newspapers. Only use these briquettes once they have completely dried, which takes about eighteen months.
- Never use petrol, oil, kerosene or other accelerants to get a fire started. They can lead to explosions or serious house fires.
- Store firewood stacked under cover in a dry but ventilated area.
- Use smaller logs instead of one large log.
- Keep the fire burning brightly and at a hot temperature. A smouldering fire produces more smoke.

- If you regularly use a wood heater, install carbon monoxide detectors indoors and keep a fire extinguisher within easy reach.
- Whether or not you have a wood heater or open fire, you should have a smoke detector. Change the battery annually, and periodically check it to make sure it is in good working order.
- Check your chimney or flue every year or two for blockages. Have it swept if necessary.
- Recycle the ash—wood ash is a good source of lime, potash and other minerals that can enrich the soil. However, it is alkaline, so limit your use of it in your garden to three or four sprinkled applications each winter, and keep it away from acid-loving plants. Ash is a great neutralising addition to a compost bin or pile, as compost has a tendency to become acidic.

STAYING COOL

While we want warm and cosy winter evenings indoors, we don't want sweltering, sleepless summer nights. Summer comfort starts with taking steps to prevent your home gaining heat in the first place, rather than trying to cool it down once it's heated up. If your home is well insulated and protected from the sun's heat, you may not need to use additional cooling.

Keeping cool in summer
- During the day, block the sun's heat by closing curtains and blinds, particularly where there's direct sunlight.

- Still air can feel 'hotter' and more uncomfortable. Install ceiling fans to circulate air.
- At night, once the air outside the house is cooler than the air inside, open curtains and blinds and windows to let the cooler air circulate. However, don't open windows while an airconditioner is operating.
- If you live in a multi-storey house, spend time on lower floors, where the air is cooler.
- Consider fitting external window blinds and shutters.
- Fill any gaps or air leaks. Just as they let heat out and draughts in during winter, they also let hot air in during summer.
- Lights, dishwashers, cooking appliances and dryers all produce heat. Avoid using them during the heat of the day. Hang clothes outside or use a fold-up airing rack instead of a clothes dryer to dry laundry.
- Plant shade trees around the house to shade both the roof and the windows.
- Put temporary shading on the north side of your house by putting up shade cloths that can be removed during winter, or by planting deciduous trees or vines on a pergola.
- Avoid putting paving outside north-facing windows as it can reflect a lot of heat and light into the house. Ground covers, lawn, low shrubs and water features can help to cool the hot air outside.

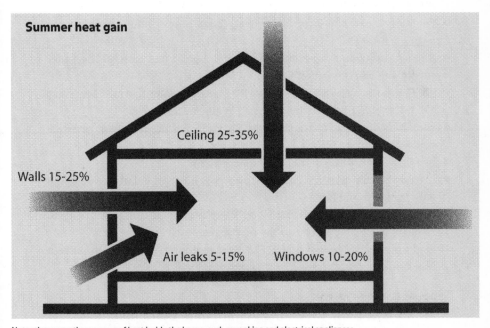

Summer heat gain

Ceiling 25-35%

Walls 15-25%

Air leaks 5-15%   Windows 10-20%

Note: there are other sources of heat inside the house, such as cooking and electrical appliances.

- Use low-emissivity (low-e) coated glass on the inside surface of the outer pane of double-glazed windows. This coating helps to keep heat in during winter and keep radiant heat from the sun out in summer.
- Now is the time for T-shirts, shorts and mini-skirts. As in winter, dress appropriately for the weather.

THE ULTIMATE COOLING GUIDE

So you've done all that you can to insulate (see chapter 12, page 235) and shade your home, yet somehow your house has turned into an oven. Perhaps you live in a warmer climate, or perhaps your home is generally comfortable, becoming sweltering for only a few weeks of the year. Either way, you may wish to install an active cooling system to get rid of the heat that you haven't been able to keep out. As with heating, decide which rooms in your house need frequent cooling. Take into account how big they are, how often they need to be actively cooled and for how long. Where possible, make sure rooms that get particularly hot can be closed off from the rest of the house.

Tips for more efficient cooling

If you do need to use additional cooling or airconditioning, keep the following in mind:

- Keep windows and doors closed while running an airconditioner.
- Close any vents in unused rooms, and close off these rooms from the rest of the house.
- Regularly clean any filters and vents. Clean outer condenser coils once a year.
- If you're using an evaporative cooler on a humid day, turn off the water supply and run the fan only.
- Set thermostats to 25–27°C (77–81°F). As with heater thermostats, every 1°C lower can increase running costs by up to 15%.
- Use airconditioning sparingly. Don't leave it running overnight or while you're out.
- Install a timer or use a programmable system.
- Use portable or personal fans wherever possible, particularly if you're the only person in the house.
- Have your ducts checked for leaks. There's no point using electricity to create cool air that leaks into the roof space.
- Find other ways to cool down. Put your feet in a cool footbath, have a cool drink or spray yourself with a refreshing mist.

Types of cooling systems

There are four types of cooling systems, all of which can provide cooling for a single room or the entire house. They generally run on electricity.

Comparing cooling systems

	FANS	EVAPORATIVE (WATER) COOLERS	AIRCONDITIONERS
Up-front costs	• $20–170 for a portable fan • $50–200 for a ceiling fan, plus installation costs	• $200–400 for portable evaporative coolers • $1000–1300 for fixed room units, plus installation • $2500–5000 for ducted central cooling systems (includes installation)	• $800–3000 for portable • $500–1600 for fixed, plus installation • $1000–10,000 for split, plus installation • $6000–15,000 for ducted central cooling (includes installation)
Running costs	Low	Relatively low	High
Greenhouse emissions	Low 0.03 tonnes CO_2/year	Relatively low 0.15–0.37 tonnes CO_2/year	Relatively high 0.86 tonnes CO_2/year for a split system 2.95 tonnes CO_2/year for a ducted system
Safety	Fitted fans need a ceiling higher than 2.7 m (around 8'1"). Remember that children can tip fans over.	Some portable units can be tipped over.	Childproof
Notes	• Fans cool by moving air; they don't reduce temperatures. They suit warm, humid climates where temperatures don't get extremely hot. • Ceiling fans can incorporate a light.	• They're more effective in areas with a dry climate and low humidity. • A window or door needs to be open so that hot outside air can be drawn in and moist air can be expelled. When considering the installation cost, factor in fitting a flywire screen to the door or window if you don't already have one. • They can be noisy inside the house, especially when the fan is on a higher setting.	• They work best if ceiling-mounted. • They're quieter than evaporative coolers inside, but can be noisy outside. • Reverse-cycle systems can be used for both heating and cooling.
Look for	• Curved blades that produce more air movement • Variable speed control • A reverse function for winter use (in ceiling fans—for more information on how this works, see page 132	• A motorised self-closing winter seal on roof-mounted ducted systems • Zone controls for ducted systems	• A high star rating (5–6 stars on the Energy Rating label—reverse-cycle models have 2 sets of stars, covering both heating and cooling modes • Inverter technology • Programmable timer and thermostat • Adjustable louvres on the vents • 'Set-back' and 'sleep' modes that adjust the thermostat setting • Zone controls for ducted systems

Electric fans have the lowest energy use and the lowest running cost and are the cheapest to purchase of all the types of cooling system. However, they do not actually make the air colder.

Evaporative cooling works by drawing in hot air from outside through a water-moistened filter. The air is cooled as it evaporates the water in the filter. However, this effect is limited by the humidity of the outside air. These coolers are best suited to dry, hot climates as they can make little difference to the temperature of already humid air. For this reason they are common in Adelaide but unusual in Queensland.

Refrigerated **airconditioners** work by taking the heat from the air indoors and moving it outside. They also remove moisture from the air. Airconditioners have the greatest capacity to cool air but also have the highest energy use, running costs and purchase price, and produce more greenhouse emissions than do other cooling systems. Central airconditioners provide cooling for the entire house. Room airconditioners can be mounted in a window or into the wall. As with room heaters, a well-placed unit may be able to provide enough cooling for more than one room. Portable models are also available. Most airconditioners have an outside component, called the compressor, through which the unwanted heat is pumped outdoors. Cooled air is released into the home through a vent or diffuser. In window- and wall-mounted units, these functions are contained in a single unit. In split systems, the outdoor component is separated from the delivery unit. Some systems can have up to three indoor delivery units for a single outdoor section. These are called 'multi-split' systems.

CFCs in airconditioners

Before throwing out an old airconditioner that is being replaced, contact your municipality to find out whether it has a hazardous waste disposal program and for advice on the right disposal method. Old airconditioners used ozone-depleting chlorofluorocarbon (CFC) as a refrigerant.

Heat pumps are basically airconditioners with a reverse function that takes heat from outside the house and uses it for winter space heating.

As for heating appliances, use the Energy Rating label information to help you choose a more efficient model.

WATER HEATING

The next energy-use big hitter is water heating. The energy use and greenhouse emissions that result from our hot water use depend on both how we heat it and how much we use. Chapter 10 is devoted to saving water. When the water you're saving is hot, there's the added bonus of saving energy (and money, for that matter). But however much water you use, you still have to heat at least some of it (unless you're a fan of cold showers), so your choice of water heating system can make a huge difference.

A hot water system is a big-ticket item, so it is tempting just to go for the cheapest option. However, you really need to choose carefully so that you have one that meets your needs, is soft on the environment and is cost-effective over its lifetime.

Hot water can account for up to a quarter of your energy bills, so it's worth getting an energy-efficient water heater. You might have to spend big to get the new system, but that's a one-off cost. Your energy bills come every couple of months. However, with solar hot water systems, the energy used is from the sun—it's clean, and it's free, at least until some evil person works out how to bill you for it! Solar energy does not contribute greenhouse gases to global warming.

What system: continuous flow or storage? electric, gas or solar?

Before you go shopping for a new hot water system, there are two important things to consider. These will greatly influence the range of products to look at. One consideration is whether the water is stored before delivery. In storage systems, water is heated and stored in an insulated tank ready for use throughout the day. Continuous flow heaters (also called instantaneous water heaters) heat water as required, so they don't need a storage tank and the hot water never runs out.

The other consideration is what energy the system uses to heat the water.

Solar hot water systems are more expensive to buy than gas or electric water heaters. However, they are cheaper to run. Sustainability Victoria estimates that solar hot water systems reduce hot water bills by more than 60% each year, resulting in a payback period of four to ten years (depending on energy tariffs). Depending on where you live, there may be government rebates on the purchase of solar hot water systems, shortening this payback period and making solar power more affordable.

In addition, solar hot water systems give you a small degree of energy independence. In the event of a disruption of electricity or gas supply (such as the Longford gas disaster in Victoria or the Auckland electricity supply failure, or even minor blackouts), you'll still be able to have a warm shower. And while your supplier sets the charges for the electricity and gas you use, no one can charge you for the use of sunshine.

Solar hot water systems are a way of using renewable energy in the form of solar thermal energy, in contrast to the electricity produced by solar photovoltaic panels. Water is circulated through a solar thermal collector— either a 'flat panel' collector or 'evacuated solar tubes'. A flat panel collector has a dark, heat-absorbing plate containing water pipes. Evacuated solar tubes have pipes passing through transparent glass outer tubes. Light rays pass through the glass and warm the pipes, while the vacuum inside the gap between the tube and the water pipe insulates the pipe, keeping the collected heat in. Evacuated tube systems are generally more efficient than flat panels and perform better on cloudy days, though they're typically more expensive.

Electric and gas water heaters, while not as green as solar heaters, have improved over recent years. Natural gas can be used for both storage and instant hot water systems. The Energy Rating label program (see page 126) covers gas hot water systems and will help you identify the most efficient models. According to Sustainability Victoria, a high-efficiency five- or six-star-rated unit can save over $50 each year on running costs, compared with a low-efficiency (and lower rated) unit.

Electric hot water storage systems can incorporate the use of cheaper off-peak electricity, making the running cost similar to natural gas. Water for the next day is heated during the night. However, you may find yourself running out of hot water if you don't have a system large enough for your household's needs, particularly a problem when you have guests to stay. Depending on the size of your home and your system, you may be able to qualify for reheating at the off-peak rate.

Conventional electric hot water systems use electrical resistance heating elements. They take longer to heat water than systems that use the combustion of fuel as the heat source. This means that they have a longer 'recovery' time—the time taken for fresh cold water taken from the mains to heat after hot water has been drained. This can be a problem if all the members of your household shower around the same time of day.

Can I go solar?

Installing a solar hot water system is a great way to reduce your household's contribution to the greenhouse effect. However, not all dwellings are suited to solar hot water systems. Your home must have certain features to effectively use a solar hot water system.

- You need a section of roof that roughly faces north (give or take an angle of 20° from true north).
- The area where the panel is to go should get direct sunlight between roughly 8 a.m. and 4 p.m. You may have to remove any shading vegetation.
- The roof needs to have a slope of 15–30°. Roofs that are steeper than 30° are dangerous to work on. Special mounting frames can produce the ideal angle but add significant cost.
- A 300-litre tank filled with water will weigh around 420 kilograms, so the roof also needs to be strong if you're considering a roof-mounted 'thermosiphon' system. Get an expert to assess the roof and its weight-bearing capacity.
- Depending on your old system, you may be able to adapt it to solar panels. The key factors are the age and type of the existing tank. Get a competent plumber with experience with solar systems to look at your existing service and advise you. A good place to find one is www.greenplumbers.com.au.

Electric instant hot water systems are available, but they generally have to use more expensive peak electricity (depending on your electricity billing plan). They are often used in flats and units where there's not enough space for a storage system or where fluing is problematic.

This doesn't mean that all electric water heaters are bad. Heat pumps, in particular, are a new kind of electricity-run storage heating system and are very efficient. They use around 65% less energy than traditional electric water heaters. However, they currently have a high purchase price of around $2500–3500, making them cost comparable to solar systems but much more expensive than other electric and gas systems. Again, savings in running costs will pay back the difference in purchase price.

What size?

The size of the hot water system you buy will depend on the type of system you choose (storage or instant) and the number of people in your house or, for instant hot water, how many outlets need to be served at once. If you have appliances that use a lot of hot water, such as a spa bath or dishwasher, then count them as an extra person. It's important to get the size right, particularly with storage systems. If it's too big, you'll be paying to heat water you're not using. If it's too small, you'll constantly be running out of hot water.

When you buy a system, consult with the supplier about the specific needs of your household and give them a brief list of all the water-using appliances in your house. They will be able to recommend a size. Also, think about whether you will have house guests often enough for this to be taken into consideration.

Weighing up your options

Choose the greenest option that suits your budget. Keep in mind that savings will help to pay back the higher purchase price of solar and electric heat pump systems. Depending on the system you choose, you may also be entitled to claim small-scale technology certificates (STCs), which can be sold, greatly offsetting some of the purchase price. STCs are covered in more detail on page 173, or visit www.orer.gov.au for more information.

The following table is a guide to the different types of hot water system and their purchase prices, running costs and greenhouse emissions.

Hot water tips

- Install your system as close as possible to where you use the hot water— the kitchen, bathroom and laundry. The further the water has to travel through pipes to get to where it's needed, the more heat is lost on the way.

FUEL	SYSTEM TYPE	PURCHASE COST (NOT INCLUDING INSTALLATION)	RUNNING COSTS	GREENHOUSE EMISSIONS*
Electricity	Off-peak storage	$600–1500	Moderately low	High
	Peak rate storage	$500–800	High	High
	Peak rate instant	$500–1100	High	High
	Heat pump	$2500–3500	Moderately low	Moderately low
Natural Gas	5 star storage	$800–1200	Moderately low	Moderately low
	2 star storage	$700–1000	Medium	Medium
	5 star instant	$800–2100	Moderately low	Moderately low
	2 star instant	$600–800	Medium	Medium
Solar	Electric boosted	$2500–4500	Low	Low
	Gas boosted	$3000–5000	Very low	Very low

*As with other comparison tables, the emissions of electric systems are based on the use of conventional electricity, as opposed to GreenPower electricity.

- When you turn the tap on, the water is cold at first. This is the water that is already sitting in the pipes between the system and the tap. If your kitchen, laundry and bathroom are not close together, then install the system nearer the kitchen, where hot water is generally used in smaller but more frequent bursts. This will help to cut down on the heat lost in the pipes and reduce the cold water wasted while waiting for the flow to warm up.
- Storage tanks can be further insulated to reduce heat loss, particularly in colder climates, with a foil-backed insulation blanket. Make sure that flues and air vents are not blocked. Also check your system's warranty to make sure that insulating the system doesn't void the warranty.
- Insulate the water pipes, particularly any outside the house.

APPLIANCES

The electricity bill of the average Australian household is about $1200 per year and increasing. This is the ongoing cost of the many things we plug into the power sockets on our walls. While it's great that washing machines and dishwashers can do some of our housework for us, and DVD players and televisions can entertain us, they don't do it free of charge!

Green living often involves a combination of the right living habits and the right technology. This section covers a number of appliance categories, providing some tips on how to use each appliance more efficiently, as well as what to look for when buying a new one.

GREEN WHITEGOODS

Whitegoods are large appliances such as fridges, freezers, dishwashers, washing machines and clothes dryers. They're 'big ticket' items. They can cost a lot of money up-front to purchase but can have a lifetime of 10–15 years if well looked after. The efficiency of the model you buy will make a huge difference to the running costs and energy use, so a cheaper model may be a false economy in the long run. Also, with large appliances, an opportunity to buy a new one only comes around every decade or so. Make a good choice, as you'll have to live with it for a while.

For all appliances, make sure you follow the manufacturer's instructions for running and looking after your machine. If you've lost your manual, call the manufacturer's customer service number (usually in the phone directory or on their website), tell them the model number and ask for their advice or a spare copy of the machine's manual. Many are now downloadable from the internet. Have the appliance serviced periodically, as this will help it to last longer and to perform as well as it did when new.

Refrigeration

Energy-saving tips

- Set the fresh food compartment to around 3–4°C (37–39°F).
- Set the freezer to a temperature between –15° and –18°C (0–5°F).
- Make sure that the seals on your fridge and freezer doors are airtight. Regularly clean up any food spilt on the seals. If the rubber has deteriorated, contact the manufacturer, as replacement seals may be available.
- Place your fridge or freezer in a cool position, away from heat-producing appliances such as a dishwasher or oven.
- Keep any exposed condenser coils (on some older models) clean and dust-free.
- If you have a second fridge or freezer, turn if off and leave the door ajar when it's not in use.
- Regularly defrost your fridge if it isn't a 'frost-free' model.
- Make sure that the back of the fridge has adequate airflow. Allow a gap of at least 8 centimetres (3.15 inches) between the back of the fridge and the wall.
- Don't put hot food straight into the fridge or freezer. Allow it to cool first.

Retire the beer fridge

A third of Australian households have two or more fridges in use, with older second fridges often kept in the garage as a 'beer fridge'. However, older models tend to have poor or damaged seals, and are much less efficient than newer versions. Don't keep a second fridge operating if you only use it occasionally. Unplug it when you're not using it or get rid of it altogether. The Australian Greenhouse Office estimates that switching off the second fridge can save up to a tonne of greenhouse gases and $120 in energy costs each year. You can always keep the drinks cold at your next party or barbecue with a bathtub full of ice.

Buying a fridge or freezer

- Look for an energy-efficient model with three and a half or four stars. (Note that the ratings for fridges and freezers have been updated recently and the standards lifted, so these models may have stickers for five- or six-star ratings under the old rating system.) It may cost more to buy up-front, but it will repay the money in reduced energy costs over its lifetime.
- In most cases, the larger the fridge is, the more energy it uses and the more it costs to run. Buy the right size for your household's needs.
- Two-door fridges with a top or bottom freezer are generally more efficient than side-by-side models.
- Avoid automatic ice-makers and drink-dispensers set into the doors, as they increase energy consumption and cost more to buy.
- Many frost-free refrigerators are less efficient. Remember to check the energy rating.

Dishwashing

Energy-saving tips

- If your dishwasher has an energy-saving, economy or 'eco' setting, use it.
- Wash only full loads.
- Regularly clean the filter.
- Use the no-heat or air-dry option on your dishwasher if you have it. Air dry loads overnight with the door ajar.

Energy-wise dishwashers

- About 80% of the energy dishwashers need to operate is used to heat water. Buying one that uses less water means less water to heat and less energy needed to heat it.
- Dishwashers built today use around 95% less energy than those built 30 years ago.

Buying a dishwasher

- Look for an energy-efficient model (four stars or higher).
- Look for models that are water-efficient, preferably four-star-rated models.
- If you have solar-heated hot water, then choose a dishwasher that has a dual water connection (that is, a connection to both hot and cold water taps).
- If you have a small household, consider buying a machine that has a half-load washing option. Or if there are only one or two of you, consider washing by hand.

Clothes washing

Tips for using washing machines

If your washing machine is in good working order, chances are you're not going to fork out several hundred dollars to buy a new greener one. However, there are a few tips that will improve the efficiency of your existing washing machine, saving energy, water and money, and probably producing a cleaner load. If you do get a new machine, these tips still apply and will help you to run your machine in a more environmentally friendly manner and maximise its life. $

- Regularly clean the filter. Washing machines can get a build-up of lint, dirt, sand, lolly wrappers and forgotten tissues. This can lead to poor performance and ultimately a machine that simply doesn't work. Some machines have a self-cleaning function. Again, check your machine's manual for instructions on how to clear the filter, and clean it regularly.

Cold water cuts emissions

Washing clothes with cold water instead of hot can save around 225 kilograms per year in greenhouse gas emissions for a single load per week.

- Wait until you have a full load before washing. If you need to wash a small load, then use a lower water or load-size setting. Remember that washing one large load uses less water and energy than two small loads.
- Wherever possible, use a cold-wash cycle, which uses up to 90% less energy than a warm- or hot-water cycle. Many detergents are now formulated to work just as well in cold water, although you may wish to pre-dissolve powder detergents.

Tips for drying clothes

- The first and foremost tip for saving energy in the laundry is to dry clothes without the use of extra electricity. Use a solar-powered clothes dryer— a clothes line! In cooler weather you can always finish off the drying by hanging clothes in front of the heater if it's already on, or with a quick turn in the tumble-dryer. $
- When the weather is wet, dry clothes on a clothes airer or drying rack placed in front of a heater. That way you use the energy that has dried your clothes a second time to warm your house. If you have a fan-forced heater, which produces hot, dry air, the slight increase in humidity from the evaporated water may be quite a relief. However, do not place articles directly on a heater or too close to naked flames, particularly with flammable synthetic fabrics.

Dryer costs

Clothes dryers cost 27 cents or more per hour to run. Drying a single load in a clothes dryer generates more than 3 kilograms of greenhouse gas emissions, or 60 black balloons' worth, while line drying generates none.

- The spin cycle of the washing machine removes excess water much more efficiently than hot, dry air in a dryer. If you must use a dryer, use it for damp (not dripping) clothes.
- Aim to dry a 'full' load—your machine instructions will recommend load sizes.

Smaller loads waste energy. Don't overfill the dryer, as there needs to be room for the warm air to circulate.

- If you have multiple loads to dry, remember fabrics will dry more efficiently and evenly when similar fabrics are dried together. For example, try to do a separate load for towels, another for bedding and table linen, and so on.
- Rather than allowing the dryer to cool, dry two loads in a row to make use of the residual heat still in the machine.
- Don't over-dry clothes, as this wastes energy and can weaken the fabric fibres. Also, don't add wet clothes to a partially dry load, as you'll over-dry some of the original load.
- Use the cool-down cycle (just air without added heat), as this uses the residual heat in the dryer to finish drying the load. It's often called the permanent-press cycle, as loads dried with cool air crease less, reducing the need for ironing.
- Clean the lint filter after each load to ensure the machine runs efficiently. Fluff and lint can quickly build up in a dryer's filter, blocking the flow of air.

Buying a new washing machine or dryer

- Look for the Energy Rating label when you're shopping for washing machines and dryers. Choose a washing machine with a rating of four stars or more. Choose a dryer with a rating of three stars or more. Recently, large load–size dryers (for loads of up to 7 kilograms) have become available, and there are several options with six-star ratings.
- Some washing machines only have a cold water tap, and heat the water internally using electricity. If your hot water system is gas- or solar-heated, then using hot water from this system will be cheaper and will have a lower greenhouse impact. Less importantly, internal water heating also makes the wash cycle take longer.
- Size matters! If you have a big family, you may prefer a larger washing machine and/or dryer. If you're one person with a small wardrobe of outfits that you wear frequently, you may prefer a machine with a

Retiring whitegoods

Your average washing machine or fridge is a bit too large to fit into the rubbish bin. If you're lucky, the people who are selling you a new appliance may be happy to take away the old one. If not, don't dump it!

Call your local council and ask for their advice on where whitegoods can be taken. Some may have a collection service or know of one locally. The refrigerants in old fridges should be properly disposed of, as they may contain ozone-depleting CFCs. Alternatively, you could contact a local appliance repairperson, who may happily pick up your old machine as a source of spare parts.

Whitegoods are worth recycling. Steel, for example, can be extracted from old appliances and made into new steel products such as food cans, car parts, building materials and even new whitegoods.

Front-loader or top-loader?

This is the first big decision when buying a washing machine. You may think that this consideration is purely about how easy it is to load and unload, or about fitting machines into small laundries. Wrong! When it comes to the environment and looking after the condition of your fabrics, the gap between front- and top-loaders is huge.

First you have to understand how the two types of machine work. A top-loading machine is like a huge bucket that you fill with enough water to immerse the garments and then add detergent. The machine agitates the load to create small currents, which move the soapy water through the fabric. This method relies largely on the chemical action of the detergent to remove soiling and so needs more detergent. This action also subjects the clothes to a lot of pull and drag, which can wear them out and pull them out of shape much faster than hand washing or a cycle in a front-loader.

Top-loading machines also use a lot more water than front-loaders. In addition, they need enough room above them to allow the lid to open, which can prevent you from mounting a dryer above the machine in a small laundry space. The only real benefit is that their cycles are faster, but then who actually sits and watches their washing machine go through its cycle? If you're like most people and leave the machine to wash while you do something else, what's the difference between half an hour and a full hour? Front-loading machines also use less energy than top-loaders of the same capacity, despite the longer operating cycle.

The front-loading washing machine is also known as the horizontal axis washing machine, because its drum rotates around a horizontal axis. Front-loaders still use the chemical action of detergents to a degree, but enlist the help of gravity to make a physical action that's like hand washing. Enough water is added to soak the fabric. As the drum turns, the fabric is lifted higher until, thanks to the force of gravity, it falls down to the bottom of the drum again with a nice squelchy thud. It's this squelchy thud that does the work. The weight of the wet fabric against itself pushes the soapy water through the fibres, removing dirt and grime without the drag of excess water. Clothes and linen are cleaned with less water and detergent, and less wear and tear on the fabric. This is particularly important with fabrics that pill easily.

small load size so that you're not wasting water and energy with each wash. Each machine will state its load capacity. Choose a machine that reflects your household's usual washing needs. For unusually large loads—for instance, when you are spring-cleaning blankets and quilts—you can always go to a laundromat. Many laundromats now have one large, double-load machine.

- Look for a machine with a range of wash-cycle options. The closer you can tailor the cycle to the needs of the items being washed, the more efficiently your machine will run. Look out for features such as a range of settings for different fabrics, variable temperatures, variable wash times, adjustable load size, a low-energy or eco-mode option and possibly a suds-save option.

COOKING

Cooking accounts for around 5% of the greenhouse gas emissions from household energy use. In some households it can be more. This may not seem like much, particularly compared with heating and car use. However, every little bit adds up. There are now over 8 million households in Australia. If each household were to cut its greenhouse contribution by just 1%, we would generate over a million fewer tonnes of greenhouse gases each year. It's worth making an effort to save energy while cooking.

The first step is to make sure your cooking appliances are energy-efficient, instead of power-hungry, inefficient models. The next step is to make sure you're using your appliances in an energy-efficient way. Even if you're not in a position to change your cooking appliances, you can still save energy and money through energy-efficient cooking.

Tips for energy-efficient cooking

- Cook or reheat small meals in smaller appliances such as toaster ovens or microwave ovens.
- Use less energy to thaw frozen food by standing it in the refrigerator overnight instead of using a microwave or oven defrost function.
- Use a toaster instead of a griller to toast bread, crumpets and muffins.
- Match the saucepan size to the size of the hotplate that you use.
- Use only enough water to cover the food when boiling. Don't waste energy heating excess water.
- Keep lids on pots and use a lower heat setting, instead of using a higher heat with the lid off. Lids help to keep the heat inside the pot.
- Use steamers and double-boiler saucepans to cook a variety of vegetables at once.
- Keep the oven door shut. When you open the oven door you let out heat, which then has to be replaced using more energy.
- With gas cooktops, make sure that the flame isn't too big for the size of the saucepan. If you see the flame extend beyond the bottom edge of the saucepan, then you're wasting energy.
- Avoid prepared meals and cook from scratch yourself. Prepared, packaged meals use more energy—for processing, pre-cooking, packaging and transportation—than those that you prepare yourself.
- If possible, use the full oven space by cooking several dishes in it at once.
- Consider using a pressure cooker for stovetop cooking. This can cut energy use by up to 50–70%.
- Keep the surfaces of your cooking appliances clean to ensure maximum heat is reflected.

Gas or electric cooktop?

As a general rule it is usually cheaper and more energy-efficient to use natural gas rather than electricity to produce heat. This goes for space heating and heating hot water, as well as cooking. However, smaller electric appliances, such as deep-fryers, electric frypans and sandwich-makers, can be more efficient because they need to heat less space and less material.

Comparing gas flames with electric cooking elements, gas is better environmentally, is generally cheaper and is easier to cook with. However, gas can also cause breathing difficulties for people with asthma or respiratory sensitivities. Electric hotplates come in various types, including ceramic, coil, solid or induction. If you choose to go electric, choose a more energy-efficient induction cooktop.

Induction cooktops

Induction cooking is a relatively new technology and possibly the way we'll cook in the future. Electricity fed to an inductor coil inside the cooktop produces a magnetic field. When a saucepan made of magnetic material is placed on the cooktop, within the coil's magnetic field, electric currents are induced in the base of the saucepan itself and this flow of current produces heat, which then heats the food. With traditional electric stoves, the saucepan is placed on a resistance element, but with induction cooking the saucepan effectively becomes the resistance element. The cooktop surface stays cool and just provides a surface for the saucepan to sit on. This cuts down on the amount of energy that is both used and lost to the surroundings. The cooktop is also easier to clean, as spillages don't get baked onto the surface.

Induction cooktops are the most energy-efficient type of electric hotplate. They are already used by many of the world's top chefs, who enjoy the fine temperature control they offer. They produce heat very quickly. Most people accidentally burn the first few dishes they cook while getting used to the new technology.

Not all types of cookware can be used with induction cooktops. The cookware must be magnetic, such as enamelled steel, cast iron and some types of glass cookware with iron-alloy base insets. Aluminium, earthenware and other non-magnetic cookware can't be used.

Choosing an oven

There are three types of oven currently available: conventional, fan-forced and microwave. Whichever type you buy, make sure you get the right size for your needs. The larger the oven, the more space has to be heated and the more energy used to reach the desired temperature. Why buy an oven big enough for a whole turkey when Christmas only comes once a year? Larger ovens are more expensive to buy, use more energy and cost more to run.

Conventional ovens have gas burners or electric elements that provide heat. Hot air rises, so these ovens tend to be hotter at the top. Conventional ovens are no longer commonly available new, as most people prefer the more energy-efficient and evenly heating fan-forced ovens.

Fan-forced (or convection) ovens incorporate a fan, which circulates the hot air evenly around the oven. The even heat allows all shelves to be used simultaneously. These ovens heat up faster, can cook food more quickly and use up to 35% less energy than conventional ovens. Some models come with a feature that allows you to turn the fan off, allowing for foods like pastry to be browned.

Self-cleaning ovens use intense heat during their cleaning program, so you would expect them to be less efficient. However, self-cleaning ovens are generally better insulated, using less energy when they're cooking. Considering that the oven will be in cooking mode regularly, but self-cleaning only a few times each year, the self-cleaning oven will have a lower net energy use than the equivalent, standard oven.

Microwave ovens use microwave radiation to heat the food directly. They are highly efficient, as they do not waste energy on heating the food containers or the oven itself. They can also cook food in a fraction of the time it takes a conventional oven. They also tend to retain the nutrients in vegetables better than boiling in water, as this can leach the water-soluble minerals from foods. Use only microwave-safe plastic, glass or ceramic containers.

Other cooking appliances

Ovens and cooktops aren't the only cooking devices in kitchens. Many people have electric kettles, frypans, rice cookers, crockpots, toasters, deep-fryers, toaster ovens, sandwich-makers and pressure cookers.

It makes little sense heating a large oven just to warm up one piece of pizza. Use smaller cooking appliances for smaller cooking needs. They are generally very energy-efficient and cheap to run. For example, an electric frypan uses around one-quarter of the energy of a conventional hotplate to cook the same meal.

Bear in mind that it takes energy and material resources to make kitchen appliances, adding to the assorted stuff our society consumes. It seems that each Mother's Day someone comes out with a new cooking gadget, like pie and hotdog makers and even electric ice shavers and pepper grinders. Only buy what you genuinely need.

LIGHTING

In Australia we spend around $100 each year per household on the energy used to provide artificial lighting. Lighting represents 7% of our home energy use and 11% of the greenhouse emissions that stem from it. This could be halved by maximising our use of daylight, choosing the right light fittings and bulbs, and remembering to turn lights off when they aren't needed.

Already, we're making huge strides in developing greener lighting, with the phase-out of old-fashioned incandescent light bulbs. This has stimulated research and development in the lighting industry, so there are now more options than ever to provide light after dark with less carbon impact and lower running costs.

Maximising daylight

Artificial light uses electricity, which costs money and generates greenhouse gases. Daylight, on the other hand, is free and clean and, as an added bonus, is much more flattering. Knocking a hole in the wall and putting in a new or bigger window isn't the only option for increasing natural light. You can also put in a skylight or a solar light tube.

A large traditional skylight is effectively a rooftop window. While skylights provide extra light, they can also allow heat to be lost in winter and gained in summer. They can also look large and imposing on a rooftop and can cost several hundred dollars to install. Because of their large size, traditional skylights cannot always be placed over the area that needs light. Modern solar light tubes, on the other hand, can be installed fairly easily to bring in daylight and transform dark and uninviting interiors.

Solar light tubes A solar light tube consists of a small rooftop dome (around 25–40 centimetres, or 10–16 inches, in diameter), a highly reflective tube that passes through the roof to the ceiling, and a light diffuser at the ceiling end of the tube. Basically, it collects a large amount of light from the outside and reflects it into the house. The benefits of solar light tubes are that they provide natural light without affecting privacy, the light they provide is free during the day, they do not produce the heat that a normal incandescent bulb or halogen lamp produces, and they do not allow the relatively high amount of heat transfer that windows and skylights permit. They take just a couple of hours to install, and most are designed to obstruct the sun's UVA and UVB rays.

Solar light tubes can be fitted with optional exhaust fans for bathrooms, toilets and kitchens. They can also be fitted with an electric light kit that allows the room to be lit at night from the same point. Solar light tubes are a great way to bring natural light to ensuites, toilets, laundries, hallways, bathrooms and other areas. Depending on their size and rooftop location, they can provide lighting at the same level as incandescent bulbs of between 75 and 300 watts.

Providing night light

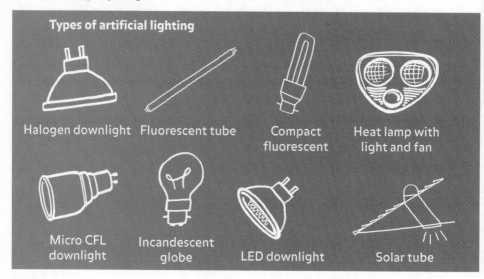

Types of artificial lighting

Halogen downlight Fluorescent tube Compact fluorescent Heat lamp with light and fan

Micro CFL downlight Incandescent globe LED downlight Solar tube

There are four types of artificial lighting that can brighten up your home at night and, more importantly, help you to see what you're doing:

- **Incandescent lighting** is provided by traditional light bulbs. Halogen lamps are also a type of incandescent lamp.
- Fluorescent tubes and the increasingly popular compact fluorescent light bulbs provide **fluorescent light**.
- The lighting technology of the future is **light-emitting diodes** or **LEDs**.
- You'll also occasionally see **solar-powered lights**, often used outdoors at a distance from a house and its electrical wiring.

Incandescent lighting In the past, incandescent lamps and bulbs were the most commonly used type of household lighting. Electricity is passed through a filament, which produces heat until it's hot enough for some of that energy to be emitted as visible light. It's a technology that's over a century old. They come in clear or pearl glass finishes, screw or bayonet fittings, and 25-, 40-, 60-, 75- and 100-watt sizes. They are cheap to buy but use a lot of electricity and only last around 1000 hours. In fact, only 4–6% of the energy provided by electricity to an incandescent bulb is converted into useful visible light. The rest of the energy is lost as heat. Despite their cheap purchase price, incandescent bulbs end up costing more through their high energy use and short life span. However, they can be used in light fittings that have a dimmer function. Most types of standard incandescent light bulb have been phased out.

Halogen lighting Halogen lamps are a type of filament light, and so produce light by producing heat first. In fact they have operating temperatures of around 200–300°C. They're commonly used as downlights, typically 50-watt dome-shaped downlights. Halogen globes are more expensive to buy than old-fashioned light bulbs, and it takes up to six downlights to provide the same general lighting as one pendant light. Halogen lights are best used for lighting a specific area, such as a painting displayed on a wall or a work area, rather than for general lighting. For safety reasons, the ceiling area around the top of each halogen light must be left free of coverings (such as insulation). A lot of halogen lights may mean a lot of holes in the insulation that keeps your home warm in winter and cool in summer. Downlights are also often vented to the roof to prevent the light from overheating. Vented downlights are a source of air leaks and can reduce the energy efficiency of a house. Downlight covers that close gaps and keep insulation away from the bulb, such as the Isolite downlight guard, are available from some specialist lighting retailers.

Recent lighting trends have seen new homes filled with halogen downlights providing general lighting. This is an enormously inefficient way to light a house. In such cases, the portion of the electricity bill that comes from lighting will be many times higher than the $100 average mentioned previously. But there are retrofit alternatives.

There are also halogen light bulbs available that look like old-fashioned incandescent bulbs and that are designed to replace them. They are often advertised as 'energy-saver halogens'. While they do use up to 30% less electricity than the equivalent conventional bulb, they don't save as much as compact fluorescent alternatives.

→ **ASK TANYA**

We are planning on building a more environmentally friendly house soon. In our current house I am trying to make some changes too. I am intending to change all our light bulbs to the modern fluorescent ones but was wondering—what do you do with the old ones? It seems very wasteful to throw working bulbs into landfill! Can they be recycled? Is there any other option?

Michelle, Northcote, Victoria

Heat lamps For most filament lamps, the heat generated is just a waste by-product, but for heat lamps the heat produced is the aim of the game. These light fittings are designed to give off radiant heat as well as light. They are commonly used as a heating source in bathrooms, in preference to forced air heaters, which have a cooling effect on damp skin. Heat lamps are often incorporated into a single bathroom unit with an alternative low-energy light and a fan, such as the IXL Tastic bathroom products. Just make sure you only use the heat lamp in winter, rather than accidentally using it as a light source in warmer months.

Fluorescent lighting Fluorescent lights are relatively expensive to buy but will more than pay back the purchase price through energy savings and their longer life. They use about a quarter of the energy used by incandescent bulbs to produce the same amount of light, and can last up to 16,000 hours.

Fluorescent tubes come in long, straight or circular tubes. They are often used in garages, workshops, kitchens and commercial and public buildings. Compact fluorescent lamps (CFLs) are more compact versions of fluorescent tubes, designed to fit into conventional light fittings. They can replace most normal incandescent bulbs.

LED lighting LED lamps have multiple light-emitting diodes as the light source—a completely different lighting technology from fluorescent or filament light technology, and one that is incredibly efficient. The technology has been used for many years in electronics, car lights and traffic signals, for example. It's also a very long-lasting and reliable technology. Options for households are starting to come on the market, with LED replacements for halogen downlights leading the way.

While it may be hard to find LED lighting in stores, and what you do find may appear expensive, keep in mind that this is an emerging technology. LEDs are rapidly improving and are coming down in price as uptake by consumers increases.

Solar lighting Solar lights have a solar panel on the top that produces electricity during the day and then stores it in a battery. This battery powers the light during the night. Once a solar light is bought, the energy that powers it is free. Another benefit is that solar lights don't have to be connected to a household electricity supply, so they are often used in gardens. Solar lights are available through hardware and gardening stores, solar equipment specialty stores and some lighting outlets.

Planning and using lighting systems

Choosing the right lighting system is just as important as choosing the right globe. General lighting provides soft light for a whole room; task lighting provides focused light over specific areas. The options are pendant lights, recessed lights, lamps and fittings with multiple globes.

Pendant lights hang down from the ceiling. They produce the most light from a single globe and so are well suited to providing general light. Lights that are recessed into the wall or ceiling are often called downlights. Downlights produce bright pools of light, rather than general lighting. It takes up to six downlights to provide as much light to a room as a single pendant light does.

Replacing light bulbs

Short of getting an electrician in, your lighting choices will be limited to the existing light fittings you have in your home.

Standard light battens come in two fitting types: screw-in and bayonet. Now that we've 'banned the bulb', the most energy-efficient replacement is the CFL, using about one-fifth of the energy of conventional incandescent bulbs. There are also halogen 'energy-saver' bulbs available, which look like ordinary bulbs and have the same screw or bayonet base but use 30% less electricity—not as great an energy saving as CFLs but they look more like the product they're replacing.

Downlight fittings also come in two types. Standard- or 'mains'-voltage fittings take 240-volt downlights with GU10 bases. Fifty-watt downlight bulbs (which have a GU10 lamp base) can be directly replaced by GU10 11-watt compact fluorescent downlights, sometimes called 'micro' or 'mini' CFLs. They cost around $15–25 per globe but last between five and seven times longer than halogens and have far lower running costs, so they pay back their purchase price.

Low-voltage fittings incorporate a transformer to convert the voltage to 12 volts. There is some energy loss associated with the transformer. They typically have MR16 bases. You can tell the difference by looking at them:

the pins on MR16 bases look like two thick pins, whereas on GU10 bases they look more like two nail heads. Note that 'low-voltage' does *not* mean low-energy or energy-saving. The energy use is indicated by the wattage.

Thirty-five-watt infrared-coated (IRC) halogen downlights can directly replace 50-watt bulbs using the same fitting. The coating reflects and concentrates infrared (heat) radiation onto the light filament so that less electricity is needed to heat the filament to the point where it produces visible light. IRC downlights still have the energy loss associated with the transformer and, like all halogen downlights, generate a lot of waste heat. They cost $5–15 per bulb and use around a third less energy.

Three-watt light-emitting diode (LED) downlights can also directly replace halogen downlights. They use a fraction of the energy and generally provide a cooler shade of light (nearer to daylight, though 'warm' and 'cool' versions are available), but they aren't as bright in appearance as halogens. Prices start at about $27 each (at time of publication), though products and their prices vary widely.

A final option is to have an electrician rewire the system to standard-voltage fittings and get rid of the transformers all together.

Lighting tips

- Remember to turn the lights off whenever a room is not being used.
- It sounds obvious, but open the curtains during the day instead of turning on lights.
- Limit your use of light fittings with multiple globes as they take more energy to produce the same lighting effect. It takes around six 25-watt incandescent globes to produce the same light output as a single 100-watt incandescent globe.

> **Need a green sparkie?**
>
> Visit the website of the EcoSmart Electricians program (www. ecosmartelectricians.com.au) to find an electrician trained and accredited in the latest in sustainable energy technology, including energy management, lighting, solar electricity systems, and energy-efficient heating and cooling.

- Using the right light bulb for the lighting level desired is more efficient and saves more energy than using a brighter bulb with a dimmer switch. Note that only certain types of CFL globe are suitable for use with dimmers. Dimmable CFLs will state their suitability on their packaging.
- Where you have a dimmer control, remember that lower (darker) settings use less electricity.
- If you like the warmer look of incandescent lighting but want to reduce your energy use, use warm-toned 'warm white' or 'deluxe warm white' CFLs.
- If you're painting a room, remember that lighter colours reflect light while darker colours absorb light.

- Choose your light fittings and lampshades carefully so that they don't block out too much of the light produced.
- Once in a while, carefully clean your fixtures and bulbs. A build-up of dirt and dust can reduce the light output by up to 50%.
- Have separate switches for each light, rather than one switch operating a series of lights. That way you can control the amount of light you have on and reduce energy wastage by over-lighting.
- Although downlights can create a beautiful mood, they use a lot of power when providing general light. Limit your use to lighting decorative features and task areas, such as a kitchen bench or reading chair.
- For front porch or outside lights, use CFLs if you want the area to be constantly lit. If you only want the light on when people are near, use a halogen light fitted with a motion detector. That way the light only goes on when it's needed.
- Use compact fluorescent lights in areas where the light stays on for long periods of time, such as the kitchen or living room. Reducing how frequently you turn the light on and off will extend the life of the globe. Use halogen or incandescent lights for rooms that are only used briefly or in fittings with a dimmer control.
- No light bulbs of any type should be put in your household recycling bin. Government waste authorities are in the process of developing fluorescent lamp recycling programs. Some states already have collection programs that generally operate through a few collection points. For advice about what to do with used CFL bulbs, contact your local council or your state or territory waste authority, or look it up online at www.recyclingnearyou.com.au. Incandescent light bulbs can be disposed of with your general household waste.

Mercury alert!

All CFLs contain a small amount of the toxic heavy metal mercury, but this is contained in the glass tube and poses no hazard so long as it is contained. Watch batteries and old thermometers contain far more. While they contain no mercury, incandescent lights running on electricity from coal-fired power plants result in more mercury entering the environment as burning fossil fuels releases mercury into the air. CFLs reduce this by using far less electricity.

What to do if a CFL breaks

Because of the mercury, broken CFL bulbs should be disposed of with caution:

- *Do not* vacuum.
- Temporarily turn off ducted heating or cooling systems. If weather permits, open windows to ventilate the room.
- Wearing gloves, pick up large pieces of glass and put them in a strong plastic bag.
- Wipe the area with a damp rag or paper towel to pick up any fine shards of glass or particles. Thoroughly clean the area with more rags or paper towels.
- For carpet areas, pick up fine particles and glass shards with large pieces of packing or masking tape.
- Put the tape, rags or paper towels in the plastic bag.
- Carefully remove the gloves and put them also in the plastic bag.
- Carefully seal the plastic bag in a second plastic bag and put the lot in the rubbish bin.

ENTERTAINING ELECTRICITY

Most living rooms are full of gadgets, gizmos and electronic devices that (theoretically) make our lives easier and entertain us. Most homes have a television and sound system. Many also have DVD players, PlayStations, electric musical equipment, electric toys and home computer systems.

All these appliances use electricity and so contribute to the greenhouse effect. Like all modern products, where the environment is concerned not all electric appliances are created equally. Some use more electricity than others. You can reduce the amount of energy your electric techno-gadgets use by carefully choosing the models you buy, changing the way you use them and occasionally choosing not to use them at all.

Fortunately, televisions now come under the Energy Rating label scheme (see page 126), so if it's a new TV you're after, choose the option that has the highest number of stars. Remember that size makes a huge difference. The larger the screen, the more power it will consume, regardless of the type of screen. Also look out for models that have backlighting provided by LED technology, but this will be reflected in the star rating. For other entertainment gadgets, look for those that are ENERGY STAR standard, but also remember that there are plenty of ways to entertain yourself other than sitting in front of the TV, computer or an electronic game.

ENERGY STAR standard

ENERGY STAR is an international standard for energy-efficient electrical equipment. Rather than rating products along a star scale from least to most efficient, the ENERGY STAR program sets a single standard and awards the ENERGY STAR label to products that meet or exceed this standard and its technical specifications. Specific standards vary from category to category. The ENERGY STAR program is supported by leading brands and covers a broad range of products, such as TVs, DVD players, stereo systems, computers, monitors, printers, scanners, multifunction devices and photocopiers.

ENERGY STAR products reduce the amount of energy consumed by using 'sleep' modes or by reducing the amount of energy needed when the product is in standby mode. They also consume less energy when in normal operating mode. This reduction of energy use helps the environment by reducing greenhouse emissions. In some cases it also reduces the amount of heat produced by the device, making it last longer. By reducing the power consumed by a product, ENERGY STAR also saves money on your electricity bills. Look for the ENERGY STAR label on products or packaging when you're buying

electronic office and entertainment equipment. For more information, see
www.energystar.gov.au.

Standby power

Have you ever got up during the night and stumbled around the house in the dark? Depending on how awake you were, you might have noticed the occasional tiny green light, or perhaps a red one blinking. These tiny lights are the sign of waiting appliances, ready for use. The convenience of having your television and stereo at the beck and call of your remote control comes at a cost to both you and the environment. The home is full of energy-using appliances, and some of them continue to use energy while on 'standby' or even when they're switched off.

By definition, standby power is the electricity consumed by electronic devices when they're not performing their primary function. It's estimated that standby power is responsible for 3% of our home energy use and 5% of home energy greenhouse emissions. This is literally hundreds of millions of dollars' worth of electricity that is *not* performing a useful task for people.

The main culprits are televisions, DVD players and set-top boxes in standby mode. The simple solution here is to turn them off manually when they're not in use. Other audio and video items also use a significant amount of energy in standby mode. Wherever you see one of those tiny lights, electricity is being used to power it. While it is only a tiny amount, it does add up over a long period of time. Computer equipment, printers and some other appliances can also use small amounts of electricity even when switched off. This power serves no function, and this wastage is often due to poor product design. The only way to prevent this is to turn the device off manually or off at the power point. You can also reduce standby wastage by avoiding poorly designed products. Electronic appliances that are ENERGY STAR compliant have limits on the amount of electricity they draw on standby.

Entertainment without electricity

The ultimate way to cut down the energy use of a television set is to turn it off and not watch it. Western societies are seeing a rise in obesity, due to a poor diet and our sedentary lifestyles. Basically, we don't get out and exercise enough! Instead of staying inside watching the TV, try some of these ideas:

- Go for a walk.
- Join a sporting club, or become a member at your local tennis courts or gym and use the facilities.
- Take Latin dancing lessons with your partner. It's fun, great exercise and great for your figure.

There will be nights that you want to spend indoors, and reading is to the mind what exercise is to the body. Exercise your brain:

- Play Scrabble—it's fun and it increases your vocabulary.
- Read a book (or even write one).
- Have a conversation. Get to know your family. When you're bored with them, invite friends over and chat with them.
- Take up yoga. As well as being good exercise, yoga helps you to relax and reduces stress levels.

Telltale signs of standby power use Certain features of electronic appliances need power to run. An appliance probably uses standby power if it has one or more of the following features:

- There is no 'off' switch.
- It has a remote control.
- It has a soft-touch keypad or controls.
- It is warm to touch near the switch when turned off.
- It charges the battery of a portable device.

To reduce standby and leak power wastage, you can do the following:

- Switch appliances off rather than leaving them on standby. Do this by turning off a master switch or power button, usually on the front of the product. If it has no master switch, turn it off at the power point.
- If the power point is hard to reach, devices can be plugged into a power board with individual switches. The power board is then kept within easier reach.
- Rechargers for mobile phones, MP3 players and other devices use standby power if plugged in, even when not connected to the device they recharge. Turn them off at the power point when not in use.
- Switch off at the wall and unplug the appliances you only use occasionally.
- Remember that some things are meant to remain on standby as part of their function. Don't turn off things like security systems, cordless telephones and timer-controlled devices.

→ **ASK TANYA**

My husband has very recently purchased two new digital TVs for our home and office use and so we now have two perfectly good analogue TVs. I have tried selling them at a nominal price but no luck, and was wondering if there is an organisation that would like them for free, or if there is any way to recycle them.

Anon.

Forget about giving old TVs away. With the combined effects of the stimulus payments and the phase-out of the analogue signal, the sale of new televisions has skyrocketed and there are piles of old ones to get rid of.

Don't dump old TVs on the doorstep of a charity or thrift store. They are hard to resell: there are always safety issues with second-hand electrical goods, and no one wants them anyway. Plus, any senior volunteers who typically staff thrift stores would risk serious injury trying to lift old TVs, particularly cathode ray tube models.

The National Association of Charitable Recycling Organisations recently did a survey of its members and found that charities typically have an oversupply of e-waste (electronic waste, including televisions), and because it costs them money to get rid of items they can't sell or redistribute, an overwhelming 95% reported that this influx of donated e-waste was actually losing money for the charities, diverting funds from important work.

RENEWABLE ENERGY

So you've dug up your old gas and electricity bills and have an idea of how much you've been using in the past; you've made an effort to reduce your electricity consumption. So what's next?

Earlier, we looked at how solar thermal energy, the free heat from the sun, can be used in solar hot water systems or to passively warm the home. We've also talked about the trade-offs when comparing electricity with natural gas. Now it's time to look at alternative sources of electricity, that wonderful flow of electrons that comes through power points to bring life to our gadgets, light bulbs and appliances.

In 2010 electricity generation contributed over 200 million tonnes of greenhouse gases to our total national greenhouse emissions. This represents over a third of our emissions. Australia has the highest greenhouse emissions per person in the world. Part of the reason for this is our high use of polluting coal to produce electricity.

Renewable energy brings an exciting range of energy alternatives. The sun doesn't stop shining, providing solar energy and moving air, which in turn provides wind energy. Hydro-power is an option as long as there's flowing water. Alternative energy gives nations the opportunity to lessen their dependence on oil and other fossil fuels.

It's important to remember that renewable energy sources, while better than fossil fuels for the health of the planet, are not without their own environmental impact. Alternative energy sources need to be chosen wisely and developed carefully so that the solution to the energy question doesn't become a problem itself. Some countries are better suited to particular energy sources because of their natural environment. For example, areas with moderately high but stable wind patterns are ideal places to put wind turbines.

The following table shows the main sources of renewable energy and how they compare.

Comparison of the main sources of renewable energy

Type	PROS	CONS
Solar	• It is limitless. • It is available all around the world (the polar regions get their supplies in bulk). • It is useful for remote areas. • It produces no greenhouse gases. • Once the initial cost of equipment is covered, ongoing power is free. • Passive solar energy can be used to heat homes, without the cost of photovoltaic (PV) cells. • Australia particularly has a lot of sunshine.	• Active solar systems using solar panels and photovoltaic (PV) cells are expensive to produce and set up. • Cloudy and overcast days can impair their performance. • Solar generators require a lot of space, which can sometimes be a problem.
Wind	• Once the initial cost of equipment is covered, ongoing power is fairly cheap. • Wind, like sunshine, won't run out. • It is useful for remote areas. • It produces no greenhouse gases. • The land used for wind farms can be used for other purposes as well.	• Wind speed can frequently change, and on some days there is no wind at all. • Windy areas tend to be on coastlines, where land value is often high. • Wind farms can be noisy. • It's believed that wind turbines can harm wildlife (particularly birds) and so can't be built near particular bird habitats. • Some people think that wind farms are an eyesore, so local tourism operators may oppose them.
Biomass (energy from plant fuels)	• Because fuel crops can be harvested and replenished, it is renewable. • Plant alcohols can be made from the waste parts of some crops and so make better use of our resources.	• It is still polluting, though less so than fossil fuels. • Land used for growing fuel crops may be better used for producing food. • Demand for land for biofuel crops may result in land-clearing and habitat loss.
Hydro	• Once the initial cost of the hydroelectric power station is covered, ongoing power is relatively inexpensive. • Water can be stored, so it can be a more reliable power source. • Less greenhouse gases and waste products are produced. • Small-scale projects can be used instead of large-scale projects to reduce the environmental impacts.	• Large dams take up a lot of land and flood previously 'dry' areas, requiring fish, animals and sometimes people to relocate. • Hydroelectric schemes can greatly upset and disturb aquatic ecosystems. In particular, they can stop the migration of fish. • They reduce the flows in these river systems, causing environmental problems downstream.
Landfill gas (the gas from rotting rubbish)	• It can be considered renewable (since we haven't yet stopped producing waste). • It captures and uses some of the greenhouse gases produced by landfills, which would otherwise contribute to the greenhouse effect. • Collection and use of the gas reduces the odours in the area of the landfill.	• Landfill gas plants are very expensive to build. • Landfill volumes will change as we change our waste-producing habits. • Much larger amounts of landfill gases are needed to produce electricity (compared with fossil fuels).
Geo-thermal	• Once the initial cost of equipment and installation is covered, ongoing power is fairly cheap. • It is renewable (since the molten core of the earth isn't going to cool down in the near future). • It doesn't require a lot of land.	• Geothermal energy is restricted to areas with geothermal activity. • There is a limit to how much steam or water can be drawn from one geothermal site.
Hydrogen	• It gives non-polluting carbon-free energy. The only by-product is water. • Hydrogen is one of the most common elements in the universe.	• It is difficult to concentrate and store.

ELECTRICITY AND GREENPOWER

Most Australian homes get their electricity through the national grid network—a network of powerlines that draws electricity from where it is produced and distributes it to the users. Electricity from both coal-fired power stations and from greener, renewable sources supplies local areas, with the remainder fed into the grid system. In the financial year 2007–08, only 6.9% of our electricity came from renewable sources. Once fed into the grid, electricity from green sources is indistinguishable from that produced by more polluting methods. However, consumer choice can directly influence the proportion of electricity from renewable sources that is fed to the grid, thanks to the GreenPower program.

GreenPower is a government-backed accreditation program that aims to drive the development of renewable energy by increasing demand. Like the vast majority of emerging technologies, green electricity costs more than the cheap electricity we get from fossil fuels, particularly in Australia, where the fossil-fuel industries have enjoyed generous government subsidies. In a sense, the GreenPower program sets up a way for the consumer to subsidise the higher costs of greener electricity.

When you choose GreenPower, your electricity retailer buys the equivalent amount of electricity you use (or a nominated percentage of it) on your behalf from approved sources such as wind, solar and hydro-power. For the price of a couple of takeaway coffees a week, you can have the satisfaction of knowing that your electricity use is less polluting, and the money you have spent on it will help to develop renewable energy in Australia.

When it comes to green electricity, the more, the better! The Australian government has set a Renewable Energy Target (RET) to ensure that 20% of our electricity comes from renewable sources by 2020. Renewable electricity bought through GreenPower is *additional* to renewable electricity counted under the RET, and electricity retailers are audited to ensure that GreenPower purchases are resulting in new investment in renewable energy.

Don't be put off by the higher cost of GreenPower. Remember that we spend $6.3 billion each year on beauty, personal care products and toiletries. Cut back spending on luxuries before you cut back spending on GreenPower. The good news is that GreenPower is cost neutral if you combine it with making your home more energy-efficient.

What to look for
Choose GreenPower electricity over conventional electricity. In most of Australia (Tasmania is a notable exception), electricity that is not GreenPower accredited is largely derived from polluting fossil-fuel sources. Changing to GreenPower is a 'no-brainer': everyone who wants to act on

climate change should be a GreenPower customer, if not generating their own electricity. For more information, visit www.greenpower.gov.au.

Don't choose non-accredited 'green electricity'. Some electricity companies market '100% renewable electricity' or 'green electricity', with no mention of the GreenPower program. In most cases, this is 'non-accredited' green electricity, which typically comes from hydroelectric schemes that have been around for decades. They offer cheaper electricity than new green electricity generators, because there is no need to cover start-up costs, such as the development of infrastructure or the purchase of new technology. Sourcing electricity from old hydro sources, companies can afford to offer customers '100% green electricity at no extra cost'. Look for the GreenPower logo in electricity advertising.

All the electricity generated by 'old renewables' is already counted towards the government's mandatory renewable energy targets, meaning that the demand for it has already been created by legislation. Consequently, consumer demand makes no difference to the existence of green electricity

from these sources. There is little net environmental benefit in choosing these products for your home or business.

Remember that price isn't the only factor. There are many companies and consumer advocacy groups campaigning on electricity prices, particularly in light of recent price increases due to things like the costs of upgrading old infrastructure, like powerline poles. There are websites and reports that list the 'best deals' on GreenPower, but these miss the whole point: that the higher cost is intended to be an investment in renewable energy. If you aim for the cheapest GreenPower product, you won't necessarily get the greenest. Find out about the different options offered by retailers servicing your area and ask about their GreenPower sources. You may wish to favour a particular source.

Go for 100% GreenPower. Don't be fooled. The label '25% GreenPower' really means 75% non-GreenPower. The remainder is either non-accredited green electricity or polluting conventional electricity. If you can afford it, choose 100% GreenPower. Again, keep in mind that you can afford to spend more per kilowatt hour of electricity if you're using less of them.

HOME RENEWABLE ENERGY SYSTEMS

While the GreenPower program ultimately links household electricity users with renewable electricity generators far removed from each other, another option for household electricity is making your own on-site. The most common types of home power systems are photovoltaic (PV) arrays (which produce solar electricity), 'micro' wind turbines and 'micro' hydro generators, which produce hydro-power from the flow of a river or large creek. Wind and hydro systems can obviously only be used in areas that are windy or that have a suitable watercourse. However, solar arrays can be used pretty much anywhere with roof space that isn't overshadowed.

Solar photovoltaic panels

Solar thermal panels in hot water systems capture the radiant heat of the sun and use it to heat water circulated through the system. In contrast, solar photovoltaic (PV) panels convert sunlight into electricity:

'Grid-connected' or 'grid-interactive' solar photovoltaic systems are, as the name suggests, connected to the electricity grid, allowing the home to draw electricity when the sun isn't shining or when its electricity demands outstrip the capacity of its system to supply it. Conversely, any electricity produced that isn't immediately needed in the house is fed onto the grid. If you use more electricity than your system produces, you pay your retailer only for the net difference. However, if you use less electricity than your system produces, you can sell that green electricity to the grid.

How to buy GreenPower

You can call the customer service number of any electricity retailer and ask about the GreenPower products they offer. They tend to give their products their own name or brand, such as 'GreenEarth Wind 100%' from Origin Energy, 'AuroraGreen' from Aurora or 'Natural Power 100%' from Synergy. Take the time to do the research and find out what is available in your state. If you have internet access, an easy way to do your research is by visiting the official GreenPower site: www.greenpower. gov.au. The GreenPower site allows you to search products by selecting your state and/or source preference. Once you've made your choice of product, simply call the retailer to make the switch. Make sure you choose a 100% GreenPower product, preferably sourced from wind or solar generators.

Many states and territories have 'feed-in tariff' schemes, which pay premiums for green electricity put onto the grid. With an adequately sized system, you may never have to pay another electricity bill.

With the price of solar photovoltaic systems falling and with government subsidies for their uptake, many households have taken up the option of making their own electricity and putting their surplus onto the grid *and* getting paid for it. The exciting news is that solar panel prices have fallen substantially in recent years. If you got a quote for a system a few years ago and found it unaffordable, it may be time to get a new quote.

'Grid-independent' or 'off-the-grid' systems store excess electricity in a battery so that it can be used in times of low output. Off-the-grid renewable energy power systems can be cost-effective energy alternatives for areas

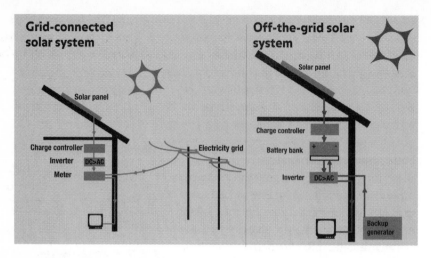

Grid-connected solar system

Solar panel

Charge controller
Inverter DC>AC
Meter

Electricity grid

Off-the-grid solar system

Solar panel

Charge controller
Battery bank + −

Inverter DC>AC

Backup generator

with high connection costs, typically rural areas without an existing connection to the grid. They are sometimes used on caravans and mobile homes.

How they work As a rule of thumb, if you can connect your system to the electricity grid, do this rather than pay the extra environmental and financial costs of a battery for storing power. As nine out of ten Australians live in urban areas, grid-connected systems are the most common installation. The electricity grid and our household appliances use alternating current electricity (AC), while solar PV panels generate direct current electricity (DC), so an inverter is needed to transform the DC electricity into AC. The inverter is connected to your home electricity meter, which reads both your electricity consumption and its production, and your meter is connected to the grid. If you have an old-fashioned meter with a spinning disc, which spins faster with high electricity use, there's something very satisfying about the first time you see the disc spinning backwards after connecting a solar system!

Location Solar panels are mostly put on rooftops. The ideal position is one with maximum 'solar access'—one that gets the most direct and concentrated sunlight for the longest period of the day, without being overshadowed. In the southern hemisphere, north-facing roofs are best; other faces will result in less electricity production. Rooftops of sheds or garages, or even systems that are mounted onto pergolas, are also an option. Panels should be angled so that the sun's rays strike them at a 90° angle.

Size The most popular systems are 1 kilowatt and 1.5 kilowatt systems, probably because of the relatively high up-front costs rather than other factors. Depending on where you live, systems in this size range produce averages of 3.5–7.5 kilowatts per day. When you consider the typical Australian household uses about 18 kilowatts per day, systems in this size range will take up to about 40% off electricity bills, a higher proportion for homes that use less electricity. Have a look at a year's worth of electricity bills and work out your average electricity consumption. Choose a size that is adequate for your needs and aims, and go for a larger system if you want to be a net producer of electricity. The table shows the Clean Energy Council's estimates of the energy-production of a 1-kilowatt solar panel system in Australia's major cities. As the output is directly proportional to the size of the system, a 2-kilowatt system (double the size) will produce roughly twice the electricity of a 1-kilowatt system, so use this table as a basis for your estimates.

Average daily production

CITY	1 KW SYSTEM
Adelaide	4.2 kWh
Alice Springs	5.0 kWh
Brisbane	4.2 kWh
Cairns	4.2 kWh
Canberra	4.3 kWh
Darwin	4.4 kWh
Hobart	3.5 kWh
Melbourne	3.6 kWh
Perth	4.4 kWh
Sydney	3.9 kWh

Source: Clean Energy Council

Costs and payback As a very rough guide, system prices start at around $9000 for a 1-kilowatt system. However, the federal government has an incentive program that awards small-scale technology certificates (STCs) for solar systems installed by professionals accredited by the Clean Energy Council. In early 2011, the Renewable Energy Target program was split into two parts: the Large-scale Renewable Energy Target and the Small-scale Renewable Energy Scheme, and it is this second scheme that awards STCs. What we now know as STCs were previously called renewable energy certificates or 'RECs'. STCs or RECs are a form of currency and can be bought, sold and traded. As such, their price is variable. Depending on the size of the system and the current trading value, STCs for a single system can add up to a few thousand dollars, making solar systems substantially more affordable.

In addition to STCs, there is also the Solar Credits scheme, which replaced the federal government solar rebate in 2009. The scheme applies a multiplier to the number of STCs you can claim. So new solar systems installed before the end of June 2012 will receive four times the number of STCs awarded based on the size of the system (a multiplier of four). The multiplier value will reduce by one each year until it runs out at the end of June 2014. It probably sounds confusing, but your system designer or installer can help with the details of your system, or you can find more information about the two schemes on the website of the Office of the Renewable Energy Regulator at www.orer.gov.au.

Separate to the STCs and Solar Credits, some states and territories have introduced a 'feed-in tariff', which pays home electricity producers a 20–60 cent per kilowatt hour premium for the cleaner electricity they're producing. In states with a net feed-in tariff, you can only claim the feed-in tariff for surplus electricity you produce over the billing period, once the total generated by the panels is calculated and the amount of electricity you've used is subtracted. Alternatively, if you're under a gross feed-in tariff, you'll be paid for every kilowatt hour of electricity you produce, whether or not it's used by your household. Householders need to sign an agreement with their electricity retailer to receive the feed-in tariff, so contact your retailer for more information and don't forget to read the fine print.

So what's the bottom line? Yes, they can cost a bit up-front, but two levels of government are offering incentives, and the prices of the panels themselves are coming down as more of them sell. Most systems pay back their purchase costs within about four years.

It's also worth mentioning the environmental costs of solar systems. They do require considerable materials and energy to be produced, but the panels themselves have a lifetime of up to 25 years. They pay back the energy needed to make them in roughly four years, though estimates vary greatly.

How to go about it

- Think carefully about what you want to achieve with the system. Do you want to produce enough electricity to cover your entire needs or just part of them?
- Think about the budget you can spend. Some banks offer low-interest loans for green technologies, so see what finance options are available.
- Gather past electricity bills. They will give you a good idea of your current level of electricity use and provide information that will help solar energy professionals tailor their advice to your needs.
- Contact several system designers or installers and get a wide range of quotes.
- Choose designers or installers that are accredited by the Clean Energy Council to ensure you get good professional advice and that your system qualifies for government incentives.
- The Clean Energy Council's *Consumer Guide to Buying Household Solar Panels* is downloadable from their site at www.cleanenergycouncil.org.au and is well worth reading before buying panels.
- Don't just find out the price of systems for which you're getting quotes; also get detailed information on the warranty and guarantee periods of the individual components and workmanship.

- Make sure you know and understand what the ongoing maintenance of the system will be, and that there are local tradespeople that can carry out any work you can't do yourself.

→ ASK TANYA

Can you please provide details on how buying solar panels and opting for 100% green electricity from an accredited supplier compare—in terms of costs to the environment?

Sally, Birkenhead, SA

We buy 100% renewable energy from our electricity supplier. Can you tell us if that is as environmentally sound as installing photovoltaic cells and going off the grid?

Premdaya, White Gum Valley, WA

These two questions are similar, so I'll answer them as one. It's hard to give a simple answer, as green electricity can come from a range of different sources (such as solar, wind, hydro and biomass) depending on where you live.

I relayed your question to Dr Mark Diesendorf, author of the book Greenhouse Solutions with Sustainable Energy *and an expert in the field of sustainable energy. He said that it's similar to asking how long a piece of string is. He made the very important point that the hefty price of solar systems may represent money better spent elsewhere. Dr Diesendorf suggested that homeowners could spend some of this money improving their house's energy efficiency, for example by excluding drafts and improving insulation. I'd also suggest having a look at the efficiency of your lighting and how well your windows are shaded in summer to keep heat out. And keep buying GreenPower.*

Don't go off the grid if you don't have to. This requires a battery, with its own environmental cost, to store the electricity. Stand-alone systems are best left for remote locations off the electricity grid network.

GreenPower or solar panels: either way you're doing something great for the planet.

Micro wind turbines and micro hydro

Micro wind turbines (MWTs) offer homes, small businesses and other organisations an alternative way to produce their own electricity, provided their property is in a windy area. Like solar PV systems, they can be grid-connected or connected to a battery for power storage, and they may also qualify for some of the same government incentives that apply for solar systems.

The big question is whether your site has adequate wind. For MWTs to be worth considering, you need average wind speeds of at least 5 metres per second (18 kilometres per hour, or 9.7 knots). If you live near a Bureau of Meteorology weather observation station, the wind speed records of that station can give you an indication of wind speeds in your area. You can view this climate data online at www.bom.gov.au/climate/data.

Once you're confident that your site is likely to have adequate wind, you will need a site assessment. This requires money and a lot of patience, but it will ensure that money isn't wasted later on a site that is poorly suited to wind power generation or in buying the wrong system. An anemometer (a device that accurately measures wind speed) will need to be set up at your site and given sufficient time (preferably a year) to measure wind speeds through the different seasons. A data-logger device records anemometer readings. Wind direction is also recorded. The site also needs to be assessed for turbulence caused by wind interacting with buildings, tall trees or other structures. Turbulent wind patterns place additional wear and tear on a turbine and so need to be minimised.

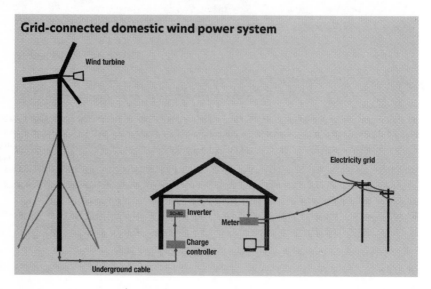

Grid-connected domestic wind power system

A number of different MWTs are now on the market in Australia, but the market for domestic wind technology is still in its infancy, just as solar PV was a decade or so ago. Currently MWT prices are very high, but in coming years they may start to come down. However, they are still likely to have relatively long financial payback periods, compared with solar PV, and they have longer energy payback periods than large-scale wind power. For these reasons, MWTs tend to be the second choice after solar PV for home renewable-energy systems. Typical buyers live in windy coastal or rural areas or for some reason are unable to use solar PV. Some renewable-energy enthusiasts opt for both.

You will also need planning permits for MWTs, and patience in obtaining them. Keep in mind that this is something fairly new for local planning authorities. Sustainability Victoria has a useful publication, the *Victorian Consumer Guide to Small Wind Turbine Generation*, which can be downloaded from www.sustainability.vic.gov.au. Information on federal government rebates and incentives is online at www.orer.gov.au.

Micro hydro systems are small hydroelectric power systems, using moving water to generate electricity. Currently, they are mostly stand-alone systems, rather than grid-connected. They're an option on properties with running rivers or streams. They have the advantage of reliability—the water flows day and night, unlike solar, and consistently, unlike wind. However, there are few places in Australia with enough water year-round, and climate change projections for Australia predict less total rainfall.

The Alternative Technology Association (www.ata.org.au) and its magazine *Renew* (www.renew.org.au) are both great resources for more information about home renewable-energy systems.

CARBON OFFSETTING

Carbon offsetting is a relatively recent method of reducing your contribution to climate change. It is one of several ways to move your lifestyle closer to being 'carbon neutral', meaning that your efforts to remove carbon from the atmosphere cancel out the amount of carbon emissions your lifestyle produces.

Carbon offsetting works by reducing or removing emissions elsewhere. Ways to do this include:

- investment in projects or technologies that save energy
- increasing the generation of renewable energy (displacing energy from fossil fuels)
- tree-planting to absorb carbon dioxide, so that carbon is stored in plant tissues.

Used appropriately as part of a broader plan for carbon neutrality, carbon offsetting is a great way to lessen our greenhouse impact. The problem with carbon offsetting is that it is misused by some companies, who see offsetting as a way to buy the right to pollute. There is also a danger that the popularity of carbon offsetting may divert much-needed funds and attention from more important efforts to improve energy efficiency.

Before considering carbon offsetting, you should reasonably exhaust your potential to improve your energy efficiency and fuel economy, and either switch to GreenPower or put solar panels on your roof. Only then should you pay an offsetting organisation to offset your net remaining emissions. Look for carbon-offsetting products that:

- invest in energy-efficiency projects
- use 100% GreenPower for all their electricity use
- plant trees and other seedlings in biodiverse plantings (providing the best habitat and local environmental benefit), and offer a guarantee for the long-term management and survival or replacement of plantings.

Carbon-offsetting organisations

As part of your efforts to become carbon neutral, you can pay the following organisations to abate carbon emissions and/or plant trees:

- **Greenfleet**—www.greenfleet.com.au
- **Climate Friendly**—www.climatefriendly.com
- **Trees for Life**—www.treesforlife.org.au.

Green gas

You may see products advertised as 'green gas' or 'green natural gas'. These are carbon-offsetting products, not alternative energy products. Like all carbon-based fuels, the use of natural gas does produce greenhouse emissions. Green gas products offer a package in which the money you pay to your gas retailer pays for both your gas use and the carbon credits needed to offset it. You can choose to do this separately with the offsetting product of your choice. However, it is possible in the future that green gas products will be offered bundled with the retailer's GreenPower products in much the same way that telecommunications companies offer discounts to people who 'bundle' their home phone, mobile phone and internet with the one provider. It's not enough for a gas company to just say that they plant a few trees. Look for independent certification of the offsetting program. For example, GreenEarth Gas from Origin Energy is independently audited.

MAKE A COMMITMENT

This week I will...

☐ _____

☐ _____

☐ _____

This month I will...

☐ _____

☐ _____

☐ _____

When I get the opportunity I will...

☐ _____

☐ _____

☐ _____

Notes

_____

_____

_____

_____

_____

_____

10

Water

There are around 1385 million cubic kilometres (1 cubic kilometre equalling 1 billion litres) of water in the world, covering nearly three-quarters of the earth's surface. This sounds like a lot of water, so you may well be wondering why there's so much fuss about conserving it and keeping it clean.

We turn the tap on; water comes out. We pull out the plug; water disappears. It's easy to take water and sewerage systems for granted when you live in the city. Very few people realise the colossal effort and investment that goes into making sure our tap water is drinkable and in good supply, our wastewater is safely treated and disposed of, and the run-off from our roofs and streets is appropriately managed. Victoria alone has $12 billion worth of water system–related assets, such as dams, pipes and treatment plants, and this doesn't include the desalination plant under construction.

Even though most of the planet is covered with water, over 97% is undrinkable salt water. A further 2% is fresh water trapped in icecaps and glaciers. More fresh water is locked too deep below the earth's surface to be extracted. What remains as available fresh water is a tiny 0.003% of all the earth's water in the forms of surface water (in lakes, reservoirs and dams, for example), accessible groundwater, soil moisture, water vapour, clouds and rain.

Although we hear of drought and the plight of farmers in news stories, the fact that we're often facing drought conditions doesn't sink in with city folk until restrictions make it an immediate reality. We tend to use water liberally, thinking that supplies are adequate as long as there are no water restrictions in place. However, in many parts of Australia water restrictions aren't put in place unless storage levels are seriously low. In 2010, eastern Australia experienced a La Niña event, with high levels of rainfall even causing flash floods and natural disasters. But Western Australia remains in drought, and it is likely that the rainfall of 2010 will provide only a temporary reprieve from drier conditions in the long term as the climate changes. We also have a growing population, which increases the demand for fresh water. To ensure that we always have enough water, everyone should continually aim to conserve water at home.

WATER ACTION PLAN

Thinking strategically, there are five basic ways that households can cut their use of water from urban water mains:

1 **Reduce** your water consumption by changing your water use habits. For example, take shorter showers and don't leave the tap running while brushing your teeth.
2 **Repair** any leaking taps, burst water mains or faulty plumbing devices. A small drip can waste 75 litres a day.
3 **Retrofit** your home with more efficient water-using fixtures, such as dual-flush toilets.
4 **Rainwater** can be collected and used instead of tap water in some instances.
5 **Recycle** your greywater.

This chapter looks in detail at where and how we can save water in our homes, reducing our consumption of mains water by changing our habits, repairing leaks, and buying and installing products and fixtures that help save water. There are also alternative sources of water that can serve some of our water needs. Recycling greywater is covered in this chapter, while rainwater collection is covered in chapter 11, pages 211–13.

WHERE WE USE WATER

The average Australian family uses hundreds of litres of water each day. This can vary greatly, depending on the size of the garden, the use of rainwater, the family itself and the climate in the area where they live. The following diagram shows where this water is generally being used, along with some estimates of the water use of particular activities.

Measure, monitor, manage

Remember that you can't manage what you don't measure. The good news is that your water bills measure your water use for you. Keep track of your water bills to monitor your progress as you make the effort to cut down your use of potable water. Remember that water use can be seasonal, so compare your current efforts with the same time last year.

Australia is the driest populated continent. Water is a precious resource that we have to conserve. With recent drought conditions, we've become increasingly water-conscious. Many homes in rural areas and some suburbs have rainwater tanks to collect and store fresh rainwater. Progress is being made in the development of home plumbing systems that recycle greywater, which is the wastewater from baths and showers, or from the washing machine, for example.

There are many things that you can do to save water and avoid pollution without installing new plumbing systems or adding to existing systems. The bonus is that saving water will save you money, especially if you reduce your use of hot water and therefore the use of the energy needed to heat it.

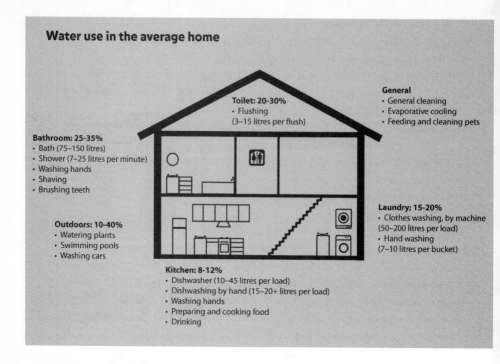

Water use in the average home

Toilet: 20-30%
- Flushing
  (3–15 litres per flush)

General
- General cleaning
- Evaporative cooling
- Feeding and cleaning pets

Bathroom: 25-35%
- Bath (75–150 litres)
- Shower (7–25 litres per minute)
- Washing hands
- Shaving
- Brushing teeth

Outdoors: 10-40%
- Watering plants
- Swimming pools
- Washing cars

Laundry; 15-20%
- Clothes washing, by machine
  (50–200 litres per load)
- Hand washing
  (7–10 litres per bucket)

Kitchen: 8-12%
- Dishwasher (10–45 litres per load)
- Dishwashing by hand (15–20+ litres per load)
- Washing hands
- Preparing and cooking food
- Drinking

Water efficiency is a lot like energy efficiency. As we've seen in the previous chapter, efficient products alone aren't enough. We also need behaviour change. Saving water in the home is a matter of combining good water-saving habits with water-saving appliances and fixtures. If you're on a limited budget, you can always start by changing your habits.

BUYING WATER-EFFICIENT PRODUCTS

If you've got a water-efficient washing machine or dishwasher, you'll be saving water every time you use it, without even trying. But how can you find water-efficient appliances? You'll find the answer in the stars—those on the Water Rating label.

The Water Rating label

Recognising the need to conserve national water supplies, the Australian government, in collaboration with state and territory governments, has introduced a Water Efficiency Labelling and Standards (WELS) scheme, which applies national mandatory water-efficiency labelling (and a minimum performance standard to toilets) to household water-using products. The WELS Water Rating label works similarly to the Energy Rating label. It also rates products on a scale of one to six stars, with a greater number of stars indicating a more water-efficient product and a better environmental choice.

The Water Rating label also states a water consumption figure. This provides an estimate of the amount of water the product typically uses when tested under Australian Standards criteria. For example, the Water Rating label for a showerhead will state water consumption as a flow rate in litres per minute, while a washing machine will state water consumption in litres per wash.

The Water Rating label can help you to choose products that will cut your water consumption and, in doing so, save a little money on your water bills. The Water Rating label takes on even more importance with devices that use hot water. High-rated water-efficient products reduce the use of water and the energy needed to heat it. So water ratings can also help you to reduce your energy costs.

Look for the Water Rating label on the following products:

- washing machines
- dishwashers
- lavatory equipment (toilets)
- showers and showerheads
- tap equipment
- urinal equipment
- flow controllers (note: labelling is optional for these).

You can also use the Water Rating search engine to compare the water ratings of available products and to calculate estimates of a product's lifetime water consumption. For more information and to use the search engine, visit www.waterrating.gov.au.

There is also the Smart Approved WaterMark program for outdoor water saving products. Read about it on page 201.

Rebates, incentives and freebies

The Australian government and the various state and territory governments, water retailers and even a handful of local governments are all offering a range of options to help householders upgrade their water-using fixtures to more efficient models. There are too many to list here, and some programs have 'use-by' dates.

The www.savewater.com.au website and education initiative provides an up-to-date list of water-efficiency rebates, incentives and special offers. Visit the 'Products' section and click on the 'Rebates and incentives' option. Examples include a $200 rebate from Sydney Water for upgrading from a single-flush toilet to a four-star dual-flush model, various levels of rebate for rainwater tanks and greywater systems, some higher rebates for rainwater tanks if they're plumbed for flushing toilets (rather than just used in the garden), or free showerhead exchange programs. You can also contact your water retailer; they will be able to provide information about their own programs as well as other state and federal schemes that apply in your region.

SAVING WATER

Why save water? The obvious reason is because we don't have an endless supply of it. But there is more to it than that. Have you ever stopped to think about where it comes from and how it gets to suburban taps?

Water is transported through tunnels and pipes from reserves, via regional storage tanks to homes. It is moved by pump, sometimes with the help of gravity. Along the way it is generally filtered, disinfected and put through a series of tests. The flow and water pressure are maintained by pressure-control valves on the mains. It takes resources and money to make sure that the water that flows through our taps is fit for human consumption. In effect, we use drinking water to wash our bodies, houses and clothes, water the garden and flush the toilet.

When we reduce our consumption of tap water, we reduce the demand on the complicated system that works to transport and purify reserves of water for the good of our health.

When it comes to saving water, do the easiest things first. Efficiency—using less water by using it wisely—is the first step. It's also the most cost-effective. Try to improve your efficiency before going to the expense of installing rainwater tanks or to the trouble of getting a greywater recycling system approved and installed. When you have an opportunity to buy or replace a water-using appliance or device, such as a washing machine or a tap fitting, look for the water conservation rating labels and choose products that use less water.

So let's take a water-saving tour of the house, room by room.

BATHROOM

Can one bathroom really make a difference? Considering that the average person has one shower or bath a day, brushes their teeth once or twice and washes their hands several times, a lot can be done in the bathroom to save water. Sometimes the smallest changes can have the biggest effect, particularly when we change things we do daily.

→ **ASK TANYA**

We want to cool our house this summer without guzzling huge amounts of electricity. I've heard that water coolers use less energy than refrigerated airconditioning. But that's adding to our water use to reduce our energy use. How much of a trade-off is it? How much water do evaporative coolers use?

Felicity, Springfield, SA

It's a cruel irony, but evaporative coolers work best in hot, dry climates where people are probably looking for ways to save water rather than more ways to use it. However, the energy savings are substantial, with evaporative coolers using around a fifth of the electricity used by the equivalent refrigerated airconditioner. Water consumption starts from about 4 litres per hour for small portable units, up to 15–30 litres per hour for ducted, whole-house systems.

Water-saving tips for the bathroom

- **Install a water-saving showerhead**. It takes one trip to a plumbing supplier or hardware shop to buy a new water-saving showerhead (or shower rose) and just a few minutes to install it. Don't accept less than a three-star rating. A standard showerhead has a flow rate of around 15 litres per minute, while a three-star-rated water-saving rose has a rate of under 9 litres per minute. In an average four-minute shower, that will save 24 litres. With one shower a day, after a year you will have saved 8760 litres of water—8784 if it's a leap year! Those showers will probably have been hot, so you'll also have saved the cost of heating all that water. Recommended showerheads are listed on the Water Rating website at www.waterrating. gov.au.

- **Install a tap-flow regulator**. Put a tap-flow regulator or an aerator (also called a 'low-flow' aerator) onto the end of your kitchen and bathroom basin taps. Both reduce the flow rate of the tap. Don't bother with an aerator for the bathtub tap as aerating the water will allow it to cool, increasing the amount of heated water you need for your bath to be the desired temperature.

- **Have a shower instead of a bath**. Since few people enjoy a shallow bath, baths use more water than quick showers, particularly if the shower has a water-saving showerhead. Save baths for when you have the time to enjoy a good soak.

- **Shave with the plug in**. Whether you're a guy shaving or a girl doing some armpit

Estimating flow rate

So you don't know the flow rate of your tap or showerhead? Just run the tap or shower into a bucket for 30 seconds. Measure how much water has collected in the bucket and double it to get the flow rate per minute for that particular fixture. Just remember to put the water in the bucket to good use. Use it on the garden or keep it by the basin for the next time you need to wash your hands.

'deforestation', don't rinse your razor under running water. A running tap releases around 10 litres of water per minute, so it's better to part-fill the sink with warm water for rinsing the razor.

- **Collect the cold water while waiting for the hot**. The cold water that's sitting in the pipe is what comes through while you're waiting for the hot water for your shower. This can be collected in a bucket and used on the garden.
- **Don't brush your teeth while the tap is running**. You only need a cup of water to rinse out your mouth, toothbrush and the basin after brushing your teeth. Use a cup when brushing your teeth instead of leaving the tap running and wasting several litres of water in the process.
- **Fix dripping taps**. Make sure you fix dripping taps straightaway, as the drips quickly add up. Periodically change the washers in your taps to prevent leaks from starting, and to make it easier to turn taps on and off properly. Alternatively, most taps can be fitted with a ceramic washer system. Ceramic washers last longer and make the taps easier to turn on and off, but don't use them in shower taps as they provide cruder adjustment of the water flow, making it harder to get the right temperature.
- **Check for leaks**. You can use your water meter to check for water leaks. At night, before going to bed, write down the meter reading. In the morning, ask the people you live with whether they've used any water during the night and, if they haven't, record the meter reading again. If the water meter reading has changed, it means there is a leak, which needs to be traced and fixed. You may need to get a plumber in.

TOILET

Nearly a third of the water we use at home is flushed down the toilet. Even in the smallest room in the house you can make a difference by saving water.

Water-saving tips in the toilet

- **Buy a top-rated dual-flush toilet**. If you need a new toilet, take note of the Water Rating label and buy a water-wise toilet. Look for models with a four-star rating.
- **Use the right button**. Make sure you use the appropriate flush button. Generally, try to use the half flush and save the big button for the big jobs!
- **Use a toilet flush regulator**. If you don't have a dual-flush toilet, you can buy a toilet-flush regulator from a hardware or

Toilet water wastage

Older toilets use around 18 litres per flush. The most efficient new toilets now use as little as 4.5 litres for a full flush and 3 litres for a 'half' flush. Let us assume an average toilet usage of one full and four half flushes (or a total of five flushes) per person each day. A person using an older-style toilet uses around 32,850 litres of water for toilet flushing each year, compared with around 6020 litres using a top-rating dual-flush toilet.

plumbing store. These devices fit inside your toilet cistern and allow you to control the amount of water used each flush by shutting off the flow of water when you remove your finger from the button. The longer you hold the flush button down, the more water will be used.

- **Improvise a half flush**. As an alternative to a flush regulator, you can simply put a water-filled plastic bottle inside the cistern (the top bit that stores the water) of an old single-flush toilet. This will reduce the capacity of the cistern by the volume of the bottle. For example, if you put in a filled 1.25-litre plastic bottle, then each time you flush you will use 1.25 litres less water.
- **Use greywater or rainwater in the toilet**. You don't need the flush water to be fresh drinking-quality water. Look into the possibility of installing a greywater recycling system so that wastewater from the shower, for example, can be used in the toilet. For more information on greywater recycling, see pages 189–92. Similarly, plumbing systems can be installed that collect rainwater from the roof, store it in a tank and pipe it to the toilet for flushing.

For the enthusiast: composting toilets

For the enthusiast: composting toilets

Composting toilets are great because they use little or no water and ultimately produce nutrient-rich fertiliser for your garden.

As the name suggests, composting toilets harness nature's ability to decompose or compost excrement and urine into harmless materials. It's a bonus that these end products can be used on the garden to make your petunias grow better.

Composting toilets generally have a fan to circulate air, which assists evaporation, speeds up the composting process and helps to prevent odours. Some also utilise the composting power of earthworms.

The drawback with composting toilets is that poorly designed or maintained systems can pose a health risk. However, there are approval processes and legislated requirements to minimise these risks. Make sure that your system complies with the relevant standards, codes of practice and regulations before having it installed.

LAUNDRY

- Wait for a full load rather than using a whole load's worth of water for just a few dirty items.
- Use water-saving options if they're available on your washing machine. Avoid extra rinse cycles unless you particularly need them—for sensitive skin, for example.
- Use green detergents that are suitable for greywater use if you plan to use your laundry water on the garden.
- Buy a top-rated washing machine. When the time eventually comes to replace your old machine, use the Water Rating label to help you choose a new one. Look for models with a four-star rating or higher. In particular, look for front-loading machines as they typically use much less water.

KITCHEN

Given that about 8–12% of the water we use in our homes is used in the kitchen, it's worth considering some ways of conserving this water:

- When washing dishes by hand, fill a second sink or a large basin with rinse water instead of rinsing dishes under running water.
- Don't use running hot water to defrost frozen food. Instead, leave it in the fridge to defrost overnight.
- Wash vegetables and fruit in a bowl or sink of water instead of under a running tap. When you're finished, you can use the wash water to water plants.
- Collect the cold water that comes out of the tap while waiting for hot water to arrive. Use this water for refilling water purifiers, watering plants, washing fruit and vegetables, and small cleaning jobs.
- Check taps for leaks and fix dripping taps.
- Install flow-regulators or aerating devices on taps. They don't cost a lot but can reduce water flow by up to 50%.
- When boiling foods in water, only use enough water to cover the food.
- Use the boiled water from steaming vegetables to make vegetable stock or, once it's cool, tip it in the garden or compost bin.
- If buying a new dishwasher, use the Water Rating label and buy a dishwasher with a rating of five stars or more.

Hand washing versus dishwasher

Dishwasher or hand washing: which uses less water? It's a question that's almost too scary to ask because we're afraid that the answer will favour hand washing, the more labour-intensive method. It all depends on how you hand wash dishes and on the model of dishwasher you're considering. Here's a run-down of the options:

- **Modern dishwashers** use around 10–45 litres of water per load, depending on their efficiency and the setting used. A four-star-rated dishwasher can use as little as 11.5 litres per load.
- **Two-drawer unit dishwashers** are available that use as little as 9 litres per drawer. Each drawer is like a half-load, giving a half-sized wash option.
- **Older-model dishwashers** vary in their water consumption, using up to 90 litres per load.
- **Dishwashing by hand**, using two sinks full of water (one for washing, one for rinsing) uses around 15–20 litres per load. It may be more for larger loads, such as after a dinner party. The amount of water used increases considerably if you rinse dishes under running tap water instead of using a filled sink or bucket.

GREYWATER

All the water that comes through our taps is of drinking quality. However, not all of our water-using activities in the home need drinkable water. 'Greywater' is the used water from the bathroom, laundry and kitchen. It is either discharged from washing machines or dishwashers, or let out down a plughole from baths, showers and basins. It doesn't include the water from the kitchen sink, toilets or bidets. Toilet wastewater is often called black or brown water, for obvious reasons.

We can reduce our demand on mains water by reusing greywater from the laundry and bathroom on the garden or for flushing the toilet. (Kitchen wastewater isn't good for reuse as it often contains a lot of fats and oils that don't break down easily and that can clog pipes. They tend to make the greywater smelly as well.) However, as well as the obvious environmental benefits of saving water by reusing greywater, there are also some environmental and health issues to consider.

Greywater contains a lot more bacteria than tap water, and these may include disease-causing organisms. It can also contain a high chemical content, depending on your use of detergents and other cleaning agents, which may be harmful to soil life and not all that good for your plants. In short, greywater reuse and recycling is not something to rush into without doing your research and getting some expert advice.

Am I allowed to use greywater?

Whether you can use greywater depends on the state or territory and council area you live in and, in some cases, whether your area is sewered or unsewered. It also depends on whether you intend to just divert greywater from the drain for immediate use or to treat and/or store the greywater in a more complex greywater system. For example, in Victoria's sewered areas you can divert greywater to your garden without a permit, provided you follow the Environment Protection Authority's list of guidelines for safe water reuse. The household reuse of greywater is a developing area for government regulation, so don't be surprised if specific regulations change over the next few years. In most states and territories, a state-level government authority approves a list of greywater recycling systems, while the local council approves the way in which a given system is used on a property. The Alternative Technology Association keeps a very useful, up-to-date listing of state and territory greywater authorities, regulations and information at www.ata.org.au (within the 'Sustainable Living Info' section).

It's important to understand the point of view of the various water and public health authorities and environmental protection agencies (EPAs). The current urban water systems were put in place to provide us with clean, safe tap water and to dispose of disease-carrying wastewater. For instance, in the nineteenth century, before Melbourne was sewered, it was affectionately known as 'Smelbourne' and disease was a major problem. This was also the case in other major cities. Good wastewater management has meant that typhoid and cholera have virtually been eliminated. Don't be surprised, then, if you don't get a lot of help or encouragement from the authorities. The water authorities have concerns about greywater recycling because if it's not done properly, things can go very wrong. They can see the need to reduce the demand on tap water, but not at the expense of public health. You may find them much more eager to talk to you about being more water-efficient or installing a rainwater tank.

General greywater requirements

Where water reuse is allowed, there are some common requirements. Legislation requires that the water must be kept within the boundaries of the property in which it was produced. It must not run off into stormwater drains or cause any sort of nuisance to your neighbours. Authorities also recommend that you limit greywater reuse in the garden to dry periods, when there is no rainwater to mix with the greywater and potentially contaminate stormwater run-off.

Installing a diversion and treatment system is a much bigger undertaking than immediately reusing greywater. You will be required to submit details of the chosen system and plans for its installation to a state authority (usually the relevant EPA) and/or local council. Local council approval processes tend to be similar to those for septic tank permits. Where government standards or guidelines aren't in place, applications are assessed on a case-by-case basis. If you would like to install a system, get advice from your water supplier. Some councils and water companies actively promote and sometimes even subsidise greywater recycling systems, particularly in dry areas, so your council may also be a valuable resource.

Any permanent changes to your plumbing must be done by a licensed or registered plumber, as required by law. Newer, greener, water-efficient plumbing products and systems are best installed by plumbers

Green Plumbers

Green Plumbers is an innovative initiative developed by the Master Plumbers' and Mechanical Services Association of Australia. The program educates and accredits interested plumbers through a series of training programs in new water-efficient plumbing technologies. Accredited plumbers are equipped with the skills and knowledge to plan and install a range of new, greener plumbing solutions. The program also provides consumers with advice and information in this area. Consultants can provide advice on solar hot water, rainwater tanks, greywater systems, government rebates, regulation requirements, and home and business water audits. To find a Green Plumber in your area, visit www.greenplumbers.com.au or call 1300 368 519.

who specialise in such technologies, such as those accredited by the Green Plumbers program. Permanent greywater diversion and treatment systems need to be well designed to make sure that they work efficiently and to ensure that the pipes carrying out the greywater and carrying in the drinking water supply can't be inadvertently cross-connected.

Ways of reusing greywater

Aside from the labour-intensive method of carting buckets of bathwater, the easiest way to reuse greywater on the garden is to access it from the drain from the laundry trough. A rubber funnel can be put through the inspection hole in the pipe with a hose attached to it. You can also use a garden hose to siphon water from the bath or laundry trough to the garden.

The next option, also for watering the garden, is to have a wastewater diverter installed by a plumber. With this device you can switch between letting the water flow down to the sewer or diverting it out through a hose to a temporary collection tank, which then disperses the water into the garden. Ideally, this tank should be connected to irrigation pipes laid in the soil a few centimetres below the surface. When the soil becomes saturated, the diverter should be switched back to the sewer.

Tips for greywater recycling

- Don't use greywater on vegetable patches or food plants where the edible portion lies on or below the ground. You can use it on plants such as fruit trees, where the part of the plant that you eat doesn't come into direct contact with the greywater.
- Ensure that there is no cross-connection of greywater with the drinking water supply. Suggest to the plumber that different coloured pipes be used for the greywater to avoid confusion in the future.
- Don't reuse kitchen water.
- Reduce the chemical content of your greywater by carefully choosing cleaning and bath products. Limit your use of bubble bath or other bath additives. In the laundry, use phosphate-free detergents. For more information about cleaning products, see chapter 4.
- If using greywater on the garden, do some watering with tap water or rainwater during drier weather. An occasional soak with rain or tap water will prevent a build-up of salt and chemicals.
- Greywater in the garden should not be allowed to pool on the surface of the ground, as this will increase the health risk. Don't overwater.
- Use a below-ground watering system or a surface drip system covered with mulch to reduce the health risk of contact with greywater and to prevent run-off.

- Use mulch on your garden as this will also reduce the health risk of contact with any excess greywater.
- Greywater can also be used outside the garden. If you're building or renovating, you can get specially designed plumbing systems that divert, treat and store greywater for use in toilet flushing. See Green Plumbers for advice.

MAKE A COMMITMENT

This week I will...

☐ _____

☐ _____

☐ _____

This month I will...

☐ _____

☐ _____

☐ _____

When I get the opportunity I will...

☐ _____

☐ _____

☐ _____

Notes

_____

_____

_____

11

The green garden

Where would a green house be without some actual greenery? A garden, whether it is a few pot plants on a windowsill or several acres, can be beautiful and can bring you a lot of enjoyment and satisfaction. Feng shui aside, it's much more pleasant and relaxing to look out of your window and see trees, shrubs and flowers than having to look at traffic, concrete and factories.

Plants and gardens put us back in touch with nature and help us to understand its processes. They help to clean the air and can even delicately perfume it. When well designed and cared for, a garden provides an attractive space for outdoor entertaining and is like an extra room. If you have children, a child-friendly garden can turn this outdoor room into an extra playroom, a place where children's imaginations can run wild, giving you some peace and quiet inside.

If you don't have a garden but want to exercise your green thumb, there are other ways to do it. Many suburbs have community gardens and allotments where fruit and vegetables are grown as a cooperative community project. There are also hundreds of 'friends of' groups (for example, Friends of Lower Kororoit Creek), Landcare groups and other tree-planting organisations that you can get involved with.

Lifestyle and aesthetics aside, a garden gives you an opportunity to compost your waste, grow your own food and support local biodiversity. All this helps the broader environment. However, there are some garden activities that can also damage the environment, through ignorant gardening.

CONSIDERATIONS FOR A GREEN GARDEN

The first thing to do is to look at the pros and cons of what you have in your garden, what you're doing with it and how it all affects the health of the planet. Consider the following lists of helpful and harmful characteristics, and work out which apply in your own garden.

How a garden can help the environment
- Trees and other plants combat the greenhouse effect and slow the effects of global warming.

- Plants soak up carbon dioxide and exhale oxygen for us to breathe.
- Trees and shrubs provide shade, keeping houses cool during hot weather. They also provide shelter, acting as powerful windbreaks.
- Plants improve water quality by filtering unwanted nutrients and pesticides.
- Native trees, plants and flowers can help conserve the diversity of local native wildlife by providing it with food and shelter.
- Garden furniture, compost bins and landscaping products can be made from recycled materials. Buying these products supports the recycling industries and leaves raw materials in the ground for future generations.
- A garden compost heap or worm farm can help to reduce the amount of food and garden waste being sent to landfill by turning it into plant food, which can be used to enrich the soil and keep your garden healthy.
- Vegetation reduces stormwater run-off. Too much stormwater run-off can cause small-scale flooding and can wash pollution into nearby waterways and the sea.
- In areas prone to the over-pumping of groundwater, filling space with vegetation instead of paving can help to prevent land subsidence. An unsealed garden bed with plants allows rainwater to seep into the soil, recharging groundwater and the water table, which supports the land above it and the structures on it.
- Vegetation helps rainwater seep into the soil, recharging groundwater supplies.
- Unsealed garden beds that soak up the rain take some of the pressure off urban stormwater systems, potentially reducing flash flooding during heavy rainfall.
- Plants can enrich the soil by converting nitrogen into nitrates.
- Vegie gardens can provide you with food. Eating locally grown produce is better for the environment and your own backyard, and growing your own food is about as local as you can get.
- Trees can prevent soil salinisation.

How a garden can harm the environment
- Gardens with plants that need more water than rainfall provides can use up a lot of fresh water.
- Pesticides can harm other non-target wildlife. Many are toxic to humans and some contain known carcinogens.
- Pesticides and artificial fertiliser can be washed into stormwater drains and carried to nearby waterways or beaches. These chemicals can harm aquatic life.
- Pesticides and artificial fertiliser can also leach into the water table and contaminate groundwater.

- Exotic (that is, non-native) plants that are unsuited to the local environment tend to require more pesticides and fertiliser.
- Introduced plant species can escape from the garden and infest nearby parks and bushlands. Some of these are prohibited under state or local laws.
- Decking, fences and other landscaping materials can be made from unsustainable plastics or rainforest timbers. Some timber products can also be treated with unsustainable and harmful chemicals, harming wildlife and potentially posing a health risk for your family.

UNDERSTANDING YOUR ECOSYSTEM

The range of plants that you can grow easily will depend on the local climate and weather patterns, the soil composition and quality, the water supply and any particular local environmental sensitivities. Each portion of the earth's surface has its own natural ecosystem. This will include certain types of vegetation that are suited to the local conditions. Planting indigenous plants (natives specifically from your local area) is an easy way to choose species that will do well in your local environment. This is not to say that non-indigenous plants will not survive in your garden. Just be aware that exotic species won't be of as much benefit to the local environment and may need more attention, water and nutrients. Plants that aren't suited to the natural landscape are high maintenance, and the land itself can suffer.

The main factors that will influence the kind of garden and plants that you can grow are outlined here:

Climate

Your local climate will nurture some kinds of plants and kill others. Tropical plants from Queensland's rainforests will struggle in the dry heat of Adelaide's summers and the frost of New Zealand winters. Rainfall patterns and leaching also influence the mineral, nutrient and salt content of the soil. Winter frosts, common in some areas, can also damage and kill plants. Remember that strong winds can uproot trees and break off branches, which can then damage houses. If you're moving to an area that gets strong winds, get a tree surgeon to check any suspect large, old trees, and ask your local nursery about more stable, deep-rooted larger trees.

Soil

The composition of soil can vary dramatically from one region to another. It can be new, more fertile soil or older, less fertile soil. It can also have varying acidity, alkalinity, and nutrient content. It can be clay-like or sandy. Since soil is basically made up of ground-up rock and decomposed organic matter or humus, your local soil will depend on the composition of the 'parent' rock and the range of plants and animals that have lived in the area. Different plants are suited to different soil types. Find out about the composition of the soil in your garden by asking your local council about soil in the area or by using a home test kit, available from hardware stores and nurseries.

Water

The next thing to consider is the supply and flow of water. You'll probably already know a little about your local rainfall patterns. Choose plants that will require little additional watering. Also consider mulch and ground covers that will help to reduce water loss, particularly in summer months. Look at any slopes within your garden. The relative heights of garden beds will affect the drainage of the soil. The root systems of some plants prefer well-drained soils. Keep this in mind when planning your garden.

Problem plants

Make sure you're aware of plants that can cause problems. Some areas have problems with seeds that escape from suburban backyards, only to germinate where they're not wanted. Most councils have lists of plants that they consider pests or that are proscribed by regional, state or local laws. Also find out about poisonous or allergy-causing plants that may pose a risk to children, pets and native wildlife. Some very common and popular garden plants fall into this group.

The final thing to consider before you start planning and planting your garden is what you want to get out of it. Think about:

- the views from various rooms in the house
- shade to help keep your home cool in summer
- shade for outdoor living and entertaining areas

- shade for pets and play areas
- positions for vegetable gardens and fruit trees
- an area for compost heaps or worm farms
- areas that can be made more private with trees and shrubs.

The rest of this chapter will explain how your choices will affect the environment. When you've finished reading, you'll be ready to look at your existing garden, decide what needs changing, make plans for a new garden and start making it happen.

WHAT TO PLANT

There are plenty of other books that cover garden design and landscaping better than this one. This chapter just covers the environmental considerations. However, an established garden does take a lot of time and will be evolving constantly.

A garden makeover that turns a weed-choked dump into the hanging gardens of Babylon in one weekend only happens on television! The advent of lifestyle television and celebrity gardeners has given us the opportunity to enjoy gardening without actually having to do any. Garden make-overs have also given people the unrealistic expectation that gardens can be transformed into a 'finished product' overnight. In reality, gardens can be changed quickly with enough money, but they are never a static, finished product. Gardens, along with areas of natural vegetation outside our fences, are constantly changing as plants grow and change size and shape, as old plants die and as new seedlings spring up. Don't expect your ideal garden to happen quickly or to stay the same once you've created it.

> **Plant size**
>
> Whenever you plant a tree or shrub, take into account the full width and height it will grow to. Make sure the plant won't grow to obstruct overhead powerlines.

Natives versus exotics

The native bush garden was once a bit of a daggy seventies trend. More recently, we've realised that this stereotype was due to the way that natives were being used, rather than the natives themselves. The bush garden is starting to become popular again, but it isn't the only way to use natives. Many gardeners have assumed that natives should be allowed to grow unrestrained, as they do in the wild. However, native plants, like any other plants, can be pruned, shaped and trained to fit into a more formal garden. In fact, most popular garden styles can be adapted to incorporate native plants. As a general rule, out of all plants, indigenous natives can bring the most benefit and least harm to your local environment.

The benefits of native plants

- Native trees provide the right habitat (that is, homes and food sources) for the wildlife of your area.
- Native trees and shrubs are better suited to the weather and soil of your area and have a better chance of thriving. Why spend your hard-earned money on an exotic that's unlikely to survive in your local conditions?
- Native plants tend to be less susceptible to local pests and plant diseases. This will reduce your need to use pesticides.
- Native plants often need less additional watering.
- Planting a tree that's native to your local area can enhance and protect the gene base of native trees in your area.

Your local council, native nursery or Landcare group can give you more information on the indigenous native trees, shrubs and grasses of your area.

Flora for fauna

Imagine that you already have a beautiful garden with soft dappled light, bright colourful flowers and lush green foliage. Now add the dimension of animal life, the colour and movement of birds and butterflies, the company of bug-eating lizards and frogs, and the sounds of birds singing. Our ever-expanding suburban sprawl has destroyed large tracts of natural vegetation that once provided habitat for native animals and birds. We can give a little back by using our gardens to provide some backyard habitat for our local wildlife.

Already, the efforts of conservation projects are restoring species that are threatened through loss of habitat. For example, the CSIRO's Richmond Birdwing Conservation Project involves households and school children in the effort to save the Richmond birdwing butterfly. This large and magnificent tropical butterfly was once common in northern New South Wales and Queensland but is now extinct in two-thirds of its original range. The

Gardens in bushfire zones

Bushfires are a frightening but normal part of nature. With global warming and an increase in the frequency of drought conditions, we're unfortunately seeing bushfires more often.

If you live in an area of high bushfire risk, you need to maintain your garden in a way that will minimise the potential damage that fire can do:

- Make sure the ground around the house is cleared of materials that can act as fuels, such as dry grass, dead leaves, branches and thick undergrowth. Trim branches well clear of the house.
- Make sure your gutters are cleared of twigs and leaf litter during high-risk periods.
- Prepare firebreak areas. For example, a well-watered lawn can act as a firebreak.
- Remove flammable items from around the house, such as wood piles, garden furniture made from wood or other flammable materials, paper recycling piles, any crates or cardboard boxes, and hanging baskets made from plant fibres.
- Make sure that you have garden hoses long enough to reach all sides of the house.
- Get a backup water supply, such as a rainwater tank with a pump.
- Consider installing garden and rooftop sprinkler systems.
- Make sure any petrol or gas stores are kept well away from the house.

community is helping to save the butterfly by gardening to help sustain its caterpillar. Residents and school children are being encouraged to remove the introduced Dutchman's pipe vine, which poisons the caterpillars, and are instead planting its food plant, the Richmond birdwing vine.

You can create some inner-urban habitat for wildlife by making your garden wildlife-friendly:

- Plant native trees and shrubs to attract and feed native birds and butterflies. For example, gardeners in the southern coastal suburbs of Adelaide can plant the native scurf pea to provide food for the caterpillar that becomes the chequered swallowtail butterfly. This butterfly has become rare in this urban area. Similarly, many areas are planting the drooping she-oak, a food source for the endangered glossy black cockatoo.
- Get a cat-proof birdbath to provide a safe water source for birds. Either hang the bath from a tree or put it on a high pedestal and regularly change the water.
- Create shelter and nesting sites for birds by placing bird boxes in trees. Surround the trees with shrubs to create a protective thicket. Birds in the garden live off seeds and small insects. They will help to keep the numbers of unwanted insects down.
- Provide hollows and sunbathing spots for lizards by partially burying pipes in garden beds, using natural mulch and including rocks and stones in the garden bed. You can place these lizard-friendly areas near your compost bin to provide the lizards with an extra food source.
- Plant nectar-producing species such as bottlebrushes and banksias to attract butterflies, insects and small birds.
- Garden ponds can be turned into frog habitats. Plant shrubs that attract insects, which provide food for the frogs. You may wish to include small native fish in the pond to prevent mosquitoes breeding in the pond.

The range of wildlife that you can attract to your garden will depend on the area you live in. Consequently, this will affect the kind of plants that you can use to create backyard habitats. Many groups and initiatives are making this information available. For more information and specific advice on what you can plant, go to the Flora for Fauna program website at www.floraforfauna.com.au.

Backyard bullies

People have often admired the romantic beauty of a willow tree, the bright-blue ready-made posy of the agapanthus, the colour and perfume of lantana or the regal simplicity of the arum lily. But put them in the wrong place at the wrong time and these beauties become garden thugs: backyard bullies that start out as the new kid on the block and end up terrorising the natives.

The biodiversity of Australia is unique, due in part to our isolation from the rest of the world's land masses. Together with many other countries around the world, Australia is facing problems caused by introduced plant species. Many of these introductions start out innocently. In their own native ecosystems, they exist in a delicate balance with other species. But once you take them to a new environment where they are free from their natural controls, such as predators or the limitations of climate, they can go feral, literally. Once established, they compete with other plants for water, soil and sunlight. As well as threatening crops and pastures, they can choke out native plants, destroying the food sources and habitat that natives provide for local animals. Some clog up rivers and waterways; others affect human health and poison animals.

There are over 2800 weeds (or unwanted plants) in Australia. About 700 of these are backyard bullies, escapees from our suburban gardens. The arum lily, for example, is now a major agricultural and environmental weed in Western Australia, covering over 10,000 hectares. Agapanthus is a favourite of lazy gardeners and local councils because it is hard to kill, and it's also a fashionable accompaniment to new Victorian- and Federation-style homes, but it has spread way beyond backyards and traffic islands in Australia's southern states and is taking over areas of native vegetation.

You may not be aware of it, but your garden could be helping the spread of backyard bullies, particularly if you live in rural areas or on the suburban fringe. You may contain the plants within your yard, but the wind can carry seeds of problem plants over great distances. Birds and other animals also eat the seeds or fruit of problem plants and spread the seeds by leaving them in their droppings.

 Contact your local council and ask about any problematic plants, particularly any popular garden plants that may be considered weeds in your area. For more information, also visit www.weeds.org.au.

What is a weed?

Don't look for a universal list of species of plants that are classified as weeds because you won't find one. Strictly speaking, any unwanted plant is a weed. Any plant in the wrong place at the wrong time can become a weed.

Controlling weeds

For the good of your health as well as that of the environment, resist the urge to nuke weeds close to home with herbicides. Herbicides can kill non-target plants as well as the weeds; they can leach into the soil and into groundwater; they can make your garden an unhealthy environment for you and your pets; and they can leave a residue on any food that you grow in your garden.

Organic weed control can be time-consuming and hard work, especially if your garden wasn't designed with weed control in mind. However, it's well worth doing. Here are some things that you can do to remove existing weeds and prevent new weed growth:

- **Clip them**. Remove the seed heads using gardening scissors or shears.
- **Dig them out**. Remove weeds by digging them out with a garden fork. Be careful not to damage the root systems of plants that you want to keep. Note that this will not be effective with species that regenerate easily from rootstock.
- **Chop them**. Use a sharp blade, spade or hoe to remove the top of the weed at ground level, without disturbing the soil.
- **Smother them**. Cover weeds with weed matting, mulch or newspaper to block out the sunlight. After a few weeks without sunlight for photosynthesis the weeds will die.
- **Solar cook them**. Cover the weeds with black plastic, particularly in hot weather. Heat will build up under the plastic and this heat will kill the weeds and their seeds.
- **Burn them**. This method also uses heat to kill the tops of weeds and their seeds, but with steam or boiling water rather than heat from the sun.
- **Pull them out**. Pulling weeds is the best, most selective way to remove unwanted weeds that have grown among plants that you want to keep. Be careful not to redistribute their seeds.
- **Give them competition**. Grow other plants and ground covers, hand-pulling the tiny weed seedlings as they appear. Once the plants you want are established, they will block the sun from new weeds and will starve the weeds of water and nutrients.
- **Block them out**. Other materials can be used in landscaping that will block out sunlight from weeds as effectively as other plants and ground covers. Consider paving or pebble gardens that will still allow rainwater to seep into the ground.

Smart Approved WaterMark

The Smart Approved WaterMark Scheme is supported and run by the Water Services Association of Australia, the Nursery and Garden Industry of Australia, the Irrigation Association of Australia and the Australian Water Association. It's Australia's water-saving labelling program for products, services and organisations that help to reduce water use outdoors and in the garden. Look for products, services and organisations that display the Smart Approved WaterMark logo.

Approved products include:
- rainwater tanks and accessories
- waterless cleaners
- mulches
- growing media
- soil moisture sensors
- greywater systems
- pool and spa products
- car-cleaning devices.

Approved services include:
- car-washing services
- sprinkler and irrigation services
- plumbers.

For a full list of approved products, services and organisations, visit www.smartwatermark.info.

WATER IN THE GARDEN

Outdoor water use represents up to 40% of the average Australian home's water use. Buckets and buckets of water are wasted on sprinkler systems turned on and forgotten, hosing down pathways instead of sweeping them, and watering plants that are too thirsty for the local conditions. A lot of the water used on the garden is better quality than it needs to be: it's fit for human consumption, but plants, soil and stormwater drains are the things that actually consume it.

Water-wise gardening tips

There are specific things you can do to save water while maintaining a lawn and garden, but you can also help by changing your habits with some every-day garden jobs:

- Use soil moisture sensors to help you recognise when your garden is thirsty.
- Mix water-storing crystals in with the soil when you're planting. They will help the soil retain water for longer after watering or rainfall.
- Use mulch in the garden to reduce evaporation. Better still, use mulch that's made from recycled materials.
- Sweep pathways with a broom rather than hosing them down.
- Install a well-designed water-efficient watering system. Look at products such as drip systems, tap timers and micro irrigation systems.
- Water late in the day in summer. Watering during the heat of the day will only cause more evaporation. Splashes of water on leaves heated by the sun can also scald delicate plants.

Lawns

- Reduce the amount of lawn area in your garden. Lawns generally require more water than other parts of the garden. Consider replacing lawns with paving or plant ground covers.
- If you can't resist having some lawn, then choose the right lawn. Ask your local nursery about water-efficient grasses, such as Santa Anna Bluegrass, Wintergreen and Greenlees Park varieties.
- Don't trim your grass to a length shorter than 2 centimetres. Trimming the grass too closely exposes the lawn to the sun, increasing the loss of water to evaporation and taking away some of the protective foliage. Longer grass puts down deeper roots and needs less water.
- If you need a new lawnmower, consider buying a mulching mower. Mulching mowers, as the name suggests, help to mulch the lawn, reducing the amount of watering it needs.

- If local water restrictions allow lawn watering, water by hand, using a hose. You're less likely to overwater by forgetting to turn off the sprinkler.
- Know when to stop. Water the lawn, but don't drown it.

Plants

- Water pot plants by hand, using a hose or watering can.
- Reduce water loss from pot plants by reducing the amount of exposed earth or potting mix. Plant ground covers, such as native violets, around trees in large pots. In smaller pots, cover the exposed earth with a layer of small pebbles or crushed polished glass.
- Group plants with similar watering needs together in your garden.
- Set up or plant windbreaks to protect delicate seedlings and to prevent the increased water evaporation caused by wind.
- Choose water-efficient native plants over thirsty exotic plants.

RECYCLING IN THE GARDEN

Recycling doesn't just belong in the kitchen: there's a whole range of materials that can be composted or mulched and used in the garden. This puts nutrients back into the soil, and helps the environment by reducing the amount of waste needing to be collected and sent to landfill.

Green waste or organic waste comprises any garden waste, food scraps and wood wastes that can be composted. You can recycle your organic waste, such as food scraps and garden clippings, by composting or using a worm farm at home. An added bonus is that many worm farms and compost bins are themselves made from recycled plastic. Buying these recycled plastic products helps to close the loop by putting the end-products of recycling to good use.

Recycling green waste is particularly significant in Australia. Soil is like humans: as a general rule, the older it gets, the less fertile it becomes.

How to spot a water-wise plant

There are some plants that are naturally able to survive dry conditions. They may be proficient at storing water or reducing water loss, or able to access water deep in the soil. There are a few tell-tale signs of a water-efficient plant that can help you to make choices for your garden without needing to be a horticulturalist:

- **Small leaves** mean a smaller surface area, which reduces water loss. Most water-efficient plants have small, tough leaves or needles.
- **Tough surfaces** on the outer layers of leaves or waxy surfaces often indicate water-efficient plants.
- **Light leaf colours** reflect sunlight rather than absorbing it. Look for light green, blue-green or grey-green foliage. Thirstier plants tend to have soft, dark green leaves.
- **Hairy leaves** act as windbreaks, reducing water loss through transpiration. Some plants have fine hairs around their pores.
- **Backbone** in a plant means that it has a tougher internal structure and is likely to be more water-efficient. Thirsty plants will wilt more easily.
- **Deep root systems** mean that a plant is likely to make better use of soil moisture.

Don't waste your waste. Instead, reuse it in the garden. Here are some ideas for common waste items that can be reused in the garden:

- Plant seeds and seedlings in milk cartons, with some large drainage holes punched in the bottom. The carton will protect the young seedling from the weather and pests, but the cartons will biodegrade as the plant grows.
- Shredded newspaper can be used as mulch over the summer months. The newspaper will reduce evaporation and lessen the amount of watering needed by your plants.
- Shredded newspaper is also great for composting, particularly if you find you need to balance a high content of 'green' food scraps with some dry 'brown' materials. Don't forget that tea bags, coffee grounds, eggshells and hair can be put into compost bins.
- Old carpet car mats are great for kneeling on while working in the garden.
- Plastic 2-litre milk bottles make great dog-poo scoopers in the garden for pet owners. Cut off the bottom of the bottle at an angle in order to make a scoop with a handle on it. The leftover plastic can act as a scraper to bring the poo into the scoop.
- Baby food jars are ideal for storing garden seeds.
- Orange bags can be used as nets to keep pests off fruit and vegetables.
- Unwanted CDs can be tied to the branches of fruit trees to scare away birds.
- An old favourite: stockings can be used to tie plants to garden stakes.

Australia is a very old land with a thin layer of topsoil, and because it's also a dry country, anything we can do to reduce water use is worth doing. Green waste can be turned into compost, which enriches and improves soil. Drier, brown organic waste, such as dry leaves and twigs, can also be used as mulch, protecting the root bed from the elements and reducing moisture loss from the soil. Mulch can prevent water run-off and stops up to 70% of gardening water being lost to evaporation.

COMPOST

In areas of natural vegetation, soil is slowly but constantly renewed through the natural breakdown of rock through erosion and the decomposition of organic matter. This organic matter includes leaf litter, dead bugs and other animals, fallen branches and overripe fruit. Bacteria in the soil and other little creatures such as earthworms break this organic matter down into rich, brown humus. Humus is the organic component of the soil, the part that makes it healthy and fertile. The minerals come from broken-down rock. Composting systems take the natural process of biological decomposition and concentrate it in a garden-based humus-making factory. By concentrating the rotting organic matter in one place, the heat produced by decomposition builds up, which speeds up the process. Compost bins allow food scraps and garden clippings to be turned quickly into plant food that nourishes the soil and helps the garden bloom.

Compost systems generally consist of a well-ventilated bin with an open bottom and a lid, which helps keep small animals from scavenging in the bin. You can also make an unenclosed compost heap in a corner of the garden, but using a bin makes it much easier to control, maintain and keep out pests.

→ **ASK TANYA**

I would like to use the cardboard toilet roll holders to sow veg seeds in, then drop the whole lot into the ground when planting out so there's no root disturbance. Do you know if the glue used is toxic?

Diane, Coonabarabran, NSW

I had a chat with a toilet tissue manufacturer—there's a fun answer to 'What do you do for a living?'—and he said that PVA is the glue used in cardboard toilet rolls. PVA is a non-toxic synthetic polymer. I also spoke with Simon Leake, Principal Soil Scientist at Sydney Environmental and Soil Laboratory. He said the PVA would rapidly degrade to harmless substances and that it wouldn't be a problem for the health of the soil. However, he did say that a cardboard toilet roll used in this way would last about a month before disintegrating, which you might like to consider when choosing what to sow. They might not last the distance for seeds that take a while to germinate.

Steps to home composting

1 **Set it up.** Compost bins are available from hardware and gardening stores. They start from $40 for a basic bin. Look for compost bins that are made from recycled plastic. You can also get compost bins on support frames that allow the bin to be rotated by turning a handle. Choose a shady spot for your compost bin because too much sunlight can dry out the heap and slow down decomposition. It needs to be well ventilated, so don't position it too close to trees, fences or walls. Try to locate it within easy access of your kitchen.

Applying compost

You don't have to break your back digging compost into the soil. Spread compost on the surface of your garden beds. Earthworms will do the job of carrying the nutrients deeper into the garden bed for you.

2 **Build the layers.** Gradually fill your compost bin with alternating layers of food scraps, garden clippings and shredded paper. An occasional layer of soil also helps. A compost heap benefits from a mix of green components such as vegie scraps and fresh lawn clippings, and brown components such as dry leaves and shredded newspaper. Blood-and-bone or dolomite can also be added to the heap to enrich the compost. You can also add small amounts of poultry or pigeon manure to give the compost extra nutrients. Do not add meat, dairy products, fish, bones, bread, rice or oily food. Do not add clippings from diseased plants or those containing seeds. Do not add dog, cat or human manure as it can spread disease.

3 **Maintain it.** Keep your compost moist and well ventilated. Occasionally turn and mix the layers with a garden fork. You can also add compost worms (available from gardening centres) to compost to help break the organic matter down faster.

4 **Use the compost.** It takes around four months for the bulk of the organic matter to break down. When it's dark and crumbly, it's ready to use. Some compost bins have doors at the base so that you can remove deeper, older

compost that may be ready while the top of the heap is still breaking down. You may wish to have two compost bins, filling one bin first, then putting fresh material into the second bin while the contents of the first bin breaks down. Dig the compost into your garden beds or spread it over the top of established beds as mulch. Remember to wear gardening gloves when handling the compost.

Compost troubleshooting

PROBLEM	SOLUTION
The compost smells bad.	The pile may be too wet, have too much green content or may not be well ventilated enough. Mix in some dry ingredients, make sure it's well ventilated and, wearing a mask, turn it with a garden fork.
The compost attracts small animals and blowflies.	Small vinegar flies are a sign of a functioning compost heap. Blowflies, rodents and other animals are often attracted to meat and dairy products, so they may be a sign that you've added the wrong ingredients. Rodents are also attracted to the warmth and insulation of a compost heap in winter. Cover each addition of food with a good layer of soil. Set rodent traps around the bin. Also consider equipping the bin with a wire mesh 'pest-proof' bottom.
The compost is taking too long to break down.	Compost systems need air, moisture and warmth to break down. Check that the pile is damp and well ventilated and that there's a balance of green and brown waste.

WORM FARMING

Earthworms are fantastic things for gardens. They do a great job of aerating and enriching the soil. Worms are proficient at munching through copious amounts of organic waste and are great helpers in reducing the amount of household waste sent to landfill. Composting using a worm farm is also known as 'vermicomposting'. Worm castings are a good natural fertiliser and soil conditioner. In its passage through the worm, the mineral subsoil undergoes changes that make the minerals available for plants. The castings contain five times the nitrate, seven times the available phosphorus, 11 times the potassium, three times the exchangeable magnesium, and 1.5 times the calcium that occur in the top 15 centimetres of uneaten soil.

However, worms don't need a garden to chomp through food scraps. They can be kept in a worm farm—a food scrap garbage-disposal unit of sorts. Worm farms do not even require a yard to be kept in, so they are a good alternative for apartments or small units where space is at a premium. In fact, worms can't survive cold temperatures, so worm farms are best kept indoors, at least for the winter months. They're a great way to compost kitchen scraps, and the worm castings are good for keeping indoor plants healthy. Worms happily munch through huge amounts of food scraps. In fact, 1 kilogram of worms can eat and recycle 1 kilogram of food every day.

Worm farms have simple structures. They consist of a series of perforated layers, stacked one on top of the other, allowing a comfortable amount of room for the worms in between. Food scraps are put into the upper layer. Worms wriggle up into this layer to eat the food, leaving their castings in lower layers. Liquid worm wee trickles down to collect in the bottom, non-perforated layer.

Worm farms, like compost bins, are available from hardware and garden stores. Some local municipalities also sell worm farms to residents at a discounted rate. Worm farms generally cost around $50–100. Alternatively, you can make your own worm farm. Live worms are sold separately. Believe it or not, worm farming can become a hobby.

Build your own worm farm

1 **Make the layers.** The layers are three stackable crates, which can be made of plastic, some kinds of wood or any other lightweight, waterproof material. You could ask your local greengrocer for three polystyrene foam fruit boxes. The bottom floor has to be solid so that it contains the liquid run-off. Ideally, the bottom floor should have a tap near the base so that the liquid can be drained. Make holes in the bottoms of the other crates or boxes.

2 **Put the layers together.** Put a brick or rock into the bottom layer. Any worms that fall into the liquid can use this rock to climb back up to the upper layers, instead of drowning in the liquid. Put on the second layer, which should be perforated, and line it with a layer of shredded newspaper

Safety tip

Compost, potting mix and fertiliser products can contain a number of substances that can harm your health. They can contain spores, fungi and other allergens, bacteria and synthetic chemicals, many of which can irritate the skin, particularly if you have scratches and small cuts on your hands, as often happens when gardening. These substances can also easily become airborne when handled. Once airborne, they can irritate the eyes and enter the lungs, sometimes causing illness. For these reasons it is important to always wear gloves and a face mask when handling compost, fertiliser and potting mix. Always open new bags of these products outside and take care with how you handle them, to minimise the amount that is dispersed into the air.

Worms like to eat:

- most fruit and vegetable scraps
- soaked and shredded pizza boxes, cardboard, paper and newspaper
- leaves
- dirt
- hair
- eggshells
- cooked potato.

Worms don't like:

- onion
- citrus fruit
- raw potato
- anything you shouldn't put into a compost bin (such as meat and dairy).

and a few handfuls of soil. This newspaper and soil provides bedding for the worms. Moisten the newspaper and soil with water, and add some worms and a small amount of food scraps. Worms don't have teeth, so make sure food scraps are finely chopped to speed up the breaking-down process. Worms don't like light, so cover the bin with some hessian or newspaper or a lid. Allow the farm to settle in for a fortnight or so, before adding some more food scraps. Over the next few weeks the worms will multiply, continue to consume the scraps and produce castings. Make sure the worms have enough to eat, but don't overfeed them. Uneaten food will just rot and cause the worm farm to smell. Worms like moisture, so make sure that the worm farm doesn't dry out. After the initial moisture, the worm farm shouldn't need additional water, provided it's kept in a cool, dark place. If it does dry out a little, spray it lightly with a mist of water.

3 **Add a layer.** Once the worm layer is half-full of worms and castings, remove the covering and place the second perforated bin on top. Put fresh bedding and food scraps into this new layer and again cover it with hessian, newspaper or a lid as before. In a week or so the worms will have moved up into the new bin, leaving behind castings in the lower level.

4 **Harvest the products.** Once worms have moved into the top of the three layers, the castings can be removed from the middle layer. This tray can be later recycled as a new top layer. Spread the castings on your garden or use in potting mixes. At any time the liquid can be drawn off from the bottom layer. Use the liquid diluted with some water as a convenient liquid fertiliser for pot plants.

Worm trivia

- There are many different types of worms. Reds, tigers and African night crawlers are just a few.
- Each worm will consume its own weight in organic waste every day.
- Compost worms are hermaphrodites: each worm is both male and female. Mature worms can fertilise or be fertilised.
- Each egg capsule contains between one and 20 young (four on average), and these young worms hatch in about 21 days.
- Worms take about 60–90 days to mature. Given the right conditions, the worm farm can double in numbers every two months.
- Worms regulate their own population.
- Worm castings hold up to nine times their weight in water, and their pH level is neutral, so they will help in releasing the maximum available nutrients and minerals into the soil. This helps to make the water-soluble nutrients in worm castings accessible for plants and their root systems.
- Worm castings contain many times the available potassium, nitrogen and phosphorus of average garden soil.
- Worm castings contain other microorganisms that will enhance plant growth and will not harm even the most delicate plants. There is never any chance of over-fertilising or burning your plants.

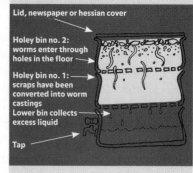

Lid, newspaper or hessian cover

Holey bin no. 2: worms enter through holes in the floor

Holey bin no. 1: scraps have been converted into worm castings

Lower bin collects excess liquid

Tap

GOING ORGANIC IN THE GARDEN

Organic gardens work with nature, rather than against it. Organic gardening involves growing plants without using artificial fertilisers and pesticides. Instead, it uses natural compost, worm castings, animal manure, green manure or a combination of these to fertilise the soil. These natural fertilisers do a much better job than their synthetic counterparts and are less likely to pollute stormwater run-off and groundwater. Diversity is the key to a successful organic garden. Many and varied plant species avoid the problems that come with plantings of a single species, such as the depletion of nutrients in the soil.

There are many benefits to organic gardening:

- You save the money otherwise spent on artificial fertilisers and pesticides.
- Your garden is a healthier environment for you, your family, any pets and local native wildlife species.
- The biodiversity of your garden's ecosystem is protected, right down to the beneficial microcultures in soil.
- Any fruit and vegetables that you might grow are healthier and often more nutritious than those bought at the supermarket.
- Organic gardens often need less watering.
- Because nature is in balance in organic gardens, once established they need very little work to maintain.
- Any run-off from your garden into stormwater drains is free from chemical pollution from pesticides and fertilisers. Polluted stormwater in our waterways and oceans can harm aquatic life.

Converting your garden to an organic garden is easier than you might think. We've already looked at garden recycling, which provides much of the nutrients your garden needs. Make your own compost or worm castings as outlined to fertilise your soil. In the meantime, if you want to add fertiliser while your compost or worm farm matures, buy organic compost mixes from your hardware or garden supplies store, and use natural pest control methods instead of chemical pesticides.

Green manure

Green manure is a way of fertilising and conditioning a patch of soil by growing a nitrogen-fixing crop, such as lucerne or beans, then digging it back into the soil. It puts nitrogen into the soil, brings up nutrients and trace elements from deeper in the soil, and adds texture to the soil.

Green manure crops are often used in crop rotation to rejuvenate the soil. The same principle can be applied to home gardens with poor soil.

Good poo, bad poo

Some kinds of animal manure are good for fertilising garden beds, and some are not.

Good poo
- poultry (most commonly chook poo)
- cow
- sheep
- goat
- horse (fibrous, so it's good for clay soils)

Bad poo
- dog
- cat
- human

MULCH

Mulch is like an insulating blanket for your garden beds. It provides a barrier between the top of the soil and the elements. Mulch reduces the loss of water through evaporation, prevents weeds from springing up, keeps the soil a stable temperature and protects the top-soil from erosion by wind and water.

As well as being functional and helpful to the environment, mulch can also be decorative. Pebbles, for example, are becoming a popular form of mulch. Mulch can be either organic, such as straw or compost, or inorganic, such as pebbles, crushed glass or gravel. Different mulches have different properties and appearances. Consider the options and choose mulch according to the needs of your garden, the cost and the look you're aiming to achieve.

Types of mulch

- **Compost** is not the most attractive of mulches, but is one of the most nutritious. It is better used on vegetable gardens than on more ornamental gardens.
- **Newspaper** makes a cheap and effective form of mulch that eventually breaks down into the soil. Modern inks are often soy-based and no longer contain heavy metals that would otherwise contaminate the soil. It can look messy and is best put on garden beds on a calm day.
- **Commercial mulch products** are readily available. Most are made from paper, wood and plant waste. They will help reduce water loss and nourish the soil as they break down. Look for those made from recycled materials.
- **Straw and hay** are low-cost mulch options. They are good for vegie gardens. Straw from legumes, such as pea straw and lucerne, adds beneficial nitrogen to the soil as it breaks down.

No-dig garden

So you're a lazy gardener: you want to start a small garden but can't be bothered doing the hard preparation and groundwork. Or perhaps the earth in your garden is more like clay or is too compacted or just poor quality. No problem! You can build a no-dig garden.

No-dig gardens are raised, boxed garden beds built directly on top of the surface of the ground. They're easy to build and save the time and effort of digging into and preparing hard earth. They're a good way to make a raised garden bed and are well suited to vegetable gardens. Because they're built on top of the soil with new organic matter, they can provide a fertile garden bed in a garden with otherwise poor soil.

First, place some edging around the proposed garden bed area. The edging can be bricks, rocks or timber. Many people use old timber railway sleepers. Spread out a thick layer (around 1.5 centimetres) of overlapping sheets of newspaper. Soak it with a hose. Then put down a 5–7-centimetre layer of lucerne (alfalfa) hay. If you have some, put down an additional layer of old leaves, twigs and pieces of seaweed. Soak with a hose. Spread a thin layer of animal manure (chicken, horse, cow or sheep) over the lucerne/twig layer. Next add a 5-centimetre layer of straw. Finally add a 3–4-centimetre layer of compost.

Plant seedlings directly into the compost layer and water them in well. Water regularly while the seedlings establish themselves. Over time the hay, straw, manure, paper and compost will break down into a dark, rich and well-aerated soil.

No-dig garden layers

| COMPOST |
| STRAW |
| MANURE |
| LUCERNE HAY |
| NEWSPAPER |

- **Leaf litter and twigs** can be gathered from your own garden and make a cheaper mulch. They can also be quite attractive, depending on the source. Pine and she-oak needle litter is particularly effective. From an aesthetic point of view, it is well suited to more wild gardens. Pine needles will also help to nourish the soil as they break down.
- **Pebbles and gravel** are a great way to cut down moisture loss in more simply styled or minimalist gardens. They go well with succulent plants and are great for covering the dirt in pot plants. They can be expensive but do not break down into the soil. Remember that if you change your mind later, they will be a pain in the neck to remove. It's worth laying some porous weed mat or mesh underneath to help with removing them, just in case. Look for human-made 'river' pebbles instead of natural pebbles and stones as the latter may have been taken unsustainably from natural river and stream habitats.
- **Crushed glass** is a recycled product used in a similar way to pebbles. The crushed glass is tumbled to remove any sharp edges. It is available in a range of colours. It can look fantastic in a well-designed garden.

RAINWATER TANKS

Why do we fail to make the connection between the water we pay for and the free stuff that falls from the sky? The reason is that it's easier for city-dwellers just to turn on the tap and use mains water than to think about installing a rainwater tank or rainwater barrel. In the country, where not everyone is connected to mains water, rainwater tanks have become a symbol of the outback, but they are not exclusive to the country. In Adelaide and its surrounding suburbs, for example, around half the houses have rainwater tanks.

Depending on the level of rainfall in your area, it's possible to use rainwater for all your water needs, but there are some issues you need to be aware of. The droplets of water that fall as rain are generally fairly clean, particularly in Australia, where we few problems with acid rain. However, when rainwater is collected it can pick up some pollutants from roof, gutter and pipe materials, bird and possum droppings, leaf litter collected on the roof, local pollution and dust. Rainwater tanks, like any still water bodies, can also breed mosquitoes. For these reasons there are regulations governing the use of rainwater tanks in Australia, and many urban councils will allow rainwater tanks for garden use or toilet flushing only.

How rainwater tanks help the environment
- Using rainwater reduces the amount of mains water used, so that our limited supplies of fresh drinking-quality water are conserved. If our

demands for fresh drinking water were to exceed the capacity of our existing reservoirs and catchments, then new dams or desalination plants would have to be built, which would have an environmental impact in particular areas.

- Rainwater tanks allow you to water your garden during times of water restrictions.
- Collecting rainwater and keeping it within your property reduces the load on your local stormwater drainage systems, which is where the run-off from your roof would otherwise flow.

Using rainwater on the garden is by far the easiest way to use it and can reduce your mains water consumption by at least a quarter. The pollutants are less of a problem for the garden than they would be if we were to drink them. Rainwater can also be used in the garden without having to change existing plumbing systems inside the house. However, you can have a plumber connect up your rainwater tank so that you can use rainwater inside the house, for toilet flushing and other uses that don't require water that's fit for human consumption. Flushing the toilet accounts for around 15–30% of the water used in the home. That means about half the average home's water use could be covered by using rainwater instead of drinking-quality tap water for gardening and toilet flushing, provided you have a large enough tank and adequate rainfall.

Stringent standards apply to rainwater systems that take water into the house. Tanks and plumbing plans have to be approved first by the relevant authorities, and professional plumbers must do the installation. Rainwater tanks are readily available from water retailers, hardware stores and specialist alternative technology stores. A 1700-litre tank will generally be adequate for a modest garden, and the force of gravity will be adequate to feed the water into a hose. Prices start from around $900. Larger sizes are available from a 2250-litre capacity (around $1000) upwards. These larger tanks can be bought with a pump or fitted with one at a later date. A pump will allow you to use the rainwater in a sprinkler system. It's worth choosing a 2250-litre or larger tank if you have space for it, particularly if you have a large garden or live in a fire-prone area and would like a back-up water supply. Buying a tank that's compatible with a pump also gives you the option of later connecting your rainwater tank for toilet flushing. Smaller 'rainwater barrels' are also available, from about $200.

Tips for collecting rainwater
- Have your roof checked to see what it's made of and painted with to make sure it's safe. Some roofs are coated with paint containing lead or petrochemical coatings, which may be toxic.

- Cover all openings with mosquito-proof mesh.
- Fit gutters and downpipes with traps to catch leaf litter and twigs.
- Install a 'first flush' diverter in the downpipe. This diverts the first few litres of rainwater into the normal stormwater drains. This water is often dirtier because it has effectively cleaned the roof of any built-up dust or pollutants.
- Make sure the tank has an overflow outlet that is connected to stormwater drains or a subsoil irrigation system.
- The tank should have air vents (covered with mosquito mesh) to prevent the water from becoming stagnant.
- Every few years, clean out the inside of the tank.
- If you live in alpine areas, you may need to partly empty the tank before winter as the stored water will expand when it freezes. This can split the sides of some tanks.

GREYWATER

Greywater, the used water from your laundry and bathroom, is another source of water for thirsty gardens. For more information on using greywater and installing a greywater recycling system, see pages 189–92. Check the requirements for reusing greywater on your garden with your local water authority or water retailer. In some states, you can reuse greywater directly on the garden. In other states you can only use greywater collected and used through a special treatment and recycling system.

PEST CONTROL

Pests are the things that you don't want to have in your garden. They may be weeds that compete with the plants you're trying to grow, bugs that eat the foliage, or diseases that damage the plants. The trouble with pesticides, apart from being a health hazard, is that they're not always very selective.

Soap, suds and soil

With regard to bathroom products, the Alternative Technology Association in partnership with RMIT University conducted a study of shampoos, conditioners, hard soaps and body washes, and of the effects of their greywaters on soil. Soil scientists were looking for signs of dispersion of clay components in soil, as this can result in drainage problems and crusting. Somewhat surprisingly, researchers found that, as a group, solid soaps had the most detrimental effect on the soil and that products that claimed to be 'organic' or 'eco-friendly' performed no better in either physical or chemical tests than their conventional counterparts.

So if you're using greywater from the shower on the garden, you might want to use a body wash, rather than a hard soap, and take claims of eco-friendliness with a grain of salt!

They can kill the plants, bugs and other organisms that you want to keep, along with those that you're trying to get rid of. When you interfere with nature, the balance between species is upset, and before you know it your garden has more problems rather than fewer. Natural pest-control methods enlist the help of nature and can be very effective. After all, cyanide occurs naturally in apricot kernels, so nature has the potential to be quite ferocious. Natural pest control makes the survival of the fittest work in favour of you and your garden.

The basics of pest control in the garden

Natural pest control and other non-chemical methods can control a range of garden pests and creepy-crawlies. As with indoor pests, there are some basic greener methods of pest control:

- **Enlist predatory allies** by encouraging the beneficial bugs and animals that eat the pests. This method lets the predators do the job of keeping pest numbers down for you.
- **Make pests unwelcome**. They will tend to leave your property if you don't allow them to establish cosy, protected nests. Be a home wrecker by carefully destroying any nests you find (as long as it's safe to do so) or by limiting the places where they might build homes, such as woodpiles or junk heaps.
- **Companion planting** works by putting plants that are like living insect repellent wherever there are plants that pests love. By planting them together you can keep the pests away from the plant you want to protect.
- **Get some chemical help**. Some natural substances are non-toxic to humans but poisonous to pests. Others simply repel them. Strategically place repellents around the garden to keep pests away or, if all else fails, use a non-toxic insecticide, such as pyrethrum.
- **Put up barriers**. Block the access of pests or birds by putting up physical barriers such as nets or wire-covered frames.
- **Use brute force**. Another option is to kill pests using old-fashioned traps, a trusty fly swat or boiling water.

Garden pest control: critter by critter

Ants aren't much of a plant problem by themselves, but they're great buddies with aphids. They love the sweet substance produced by aphids, so they protect them and carry them from plant to plant. Sprinkle the ground with bone meal to keep ants away, and plant vulnerable seedlings in sawn-off milk cartons to protect them. Geraniums, southernwood and pennyroyal plants deter ants, so plant them near aphid-prone plants such as roses and near the doorways and windows of your house.

Aphids can be removed by splashing soapy water onto the affected plant, followed by cold, clean water. Collect the soapy wash water from your laundry washing machine to do the job. Protect plants troubled by aphids by companion planting them with orange nasturtiums, catnip, catmint and garlic. Encourage ladybirds (ladybugs), as they eat a whole range of garden pests, including aphids. A single ladybird can eat 400 aphids in one week. They can be bought from some nurseries.

Birds can be both a gardener's friend and foe. While they can make a mess of seedlings, you should encourage them nevertheless, as they eat a lot of common garden pests. Make your garden bird-friendly so that you can enlist their help in pest control and enjoy their songs and company. Protect fruit and vegetables with mesh orange bags, and hang unwanted computer CDs in any fruit trees to deter the birds from eating the fruit.

Caterpillars—if they are the kind that turn into butterflies—might be something you want to encourage. To get rid of the other kinds, sprinkle finely ground pepper onto dampened plants that are threatened by caterpillars.

Earwigs like hiding places, so a very tidy garden might deter them. If you have no luck, then a linseed oil trap is very effective at killing them. Pour about 3–4 millimetres of linseed oil inside a takeaway container, put the lid on and punch some holes in the side above the surface of the linseed oil. Leave it in the garden near any infestation.

> **Make your own bug spray**
>
> Mince five cloves of garlic and around 18 hot chilli peppers. Mix in around a litre of water and allow to stand for a week or so. Strain the spicy water and put it into a labelled spray bottle.
>
> Use this spicy spray to kill ants, spiders, slugs, caterpillars and other pests. However, make sure you wash your hands well after making and using it, and be careful not to get any spray in your eyes or the eyes of pets. It can sting you as well as the pests.

Flies can be deterred by basil, tansy, pyrethrum and eau-de-cologne mint plants. Strategically place pots of these plants and citronella candles around the house near doors, windows and barbecue areas. However, keep tansy in small pots, as it can be an invasive plant.

Mealy bug populations can be diminished by ladybirds, so encourage ladybirds.

Mosquitoes can be controlled by taking away the pools of water they breed in. Don't allow such pools of water to collect outside. Keep mosquitoes outdoors by placing pots of tansy or southernwood near windowsills and doorways.

Slugs and snails have delicate bodies and don't like moving over rough ground. Sprinkling lime or wood ash around garden beds will deter them. They also don't like crawling over bark, so bark can be a good slug-resistant ground cover or mulch. Birds also eat them. Slugs and snails are killed by salt, so if deterrents or predators don't solve your slug problem, then sprinkle salt on your paths and around garden beds. Alternatively, you can set traps for them by leaving out small saucers or partially buried containers of beer. Snails and slugs love beer and will be attracted to it. Once they fall into the pool of beer, they'll drown, but at least they'll die happy.

GARDEN DECOR

A gorgeous garden consists of more than just plants. Decking, garden furniture, water features, barbecue areas and play areas for kids can turn a garden into an outdoor room, a place that you'll want to spend time in. However, some garden hardware is definitely greener than others. Like everything else, it's important that your fencing, garden furniture and other landscaping structures are made from sustainable materials that provide a healthy environment for both you and the planet. Take into account the following to make your garden more than just green to look at.

Use the good wood

It's easy to think that anything made from wood is good because it comes from a tree and we all love trees. However, sometimes using wood is bad for the environment, depending on where the wood is from and how it has been treated. As with indoor furniture, choose fencing and outdoor furniture that is made from forest-friendly timber, grown on sustainably managed plantations. Rainforest timbers may look beautiful in outdoor furniture, but they look even more beautiful when they remain rainforest trees, complete with resident wildlife. For a list of timbers to avoid, see page 234. You can also look at second-hand outdoor furniture at garage sales and second-hand shops.

Take care with wood treatments (or mistreatments)

Many outdoor wood products, including decking, furniture and play equipment, are treated with chromated copper arsenate (CCA), a highly toxic compound. This is sometimes labelled as 'Tanalised' timber or 'pressure-treated' timber. CCA is a preservative, but it has also been linked to various forms of cancer. Traces of CCA are left on the skin when the wood is handled and can also leach from wood into surrounding soil. Although studies show that this is not an immediate health threat, we still don't fully understand the cumulative effects of the many chemicals we are exposed to today. The potential for harm from CCA dramatically increases when wood treated with it is burnt, so you should never burn CCA-treated timber or other treated timbers with unknown histories. It is much safer to choose wood products that are either untreated or treated with preservatives that have a low toxicity and are arsenic-free.

You can protect wood where it meets the soil by encasing it in a metal shoe or concrete. Never treat wood posts with used motor oil, as is sometimes suggested. Used motor oil contains a cocktail of chemical nasties that can contaminate soil, groundwater, other waterways and the air. It should be safely disposed of or recycled as outlined in chapter 13.

→ **ASK TANYA**

We have some raised garden beds that we made from recycled old treated pine timber. The timber is 15–20 years old. As it was weathered, we thought that it would not be toxic. We also lined the inside surface with weed mat. Now we are concerned that perhaps this set-up is still leaching toxin. How long does the toxin stay in treated pine? Should we replace the beds with untreated oak or reline it with something else?

Steph, Seven Mile Beach, Tasmania

The Australian Pesticides and Veterinary Medicines Authority's chemical review program investigated the use of arsenic-based timber treatments, namely copper chrome arsenate (CCA) and arsenic trioxide. You can view full responses to frequently asked questions about this review online at www.apvma.gov. au/products/review/completed/arsenic_faq.php.

With respect to your question, the review says, 'CCA treatment protects pine for as long as 20 years', so it's likely to still contain some of the pesticide in question. The review also says that arsenic may leach into the soil, but how much depends on a number of factors, as does the degree of uptake by plants. The conclusion for vegie gardens was that 'in most situations the affected zone of soil is likely to be very limited, and plants growing near CCA treated posts did not have elevated levels of arsenic', but the review advised a plastic lining as a precaution. A rot-proof alternative you might like to consider is 'plastic lumber' made from recycled plastic.

Choose greener deck washes and finishes

Timber decks are traditionally sealed with flammable, petrochemically derived stains and varnishes. These sealants are not sustainable and emit air-polluting gases, which can trigger allergies. Instead, apply a deck sealant formulated from plant oils to a well-cleaned and sanded deck. Good preparation of the decking wood will ensure that the sealant will apply better and last longer.

Consider wood alternatives

You may wish to avoid wood altogether. Synthetic wood products or 'plastic lumber', such as ModWood, made from recycled plastics or recycled plastic and wood composites, are now available, though they can be harder to find. They are manufactured to need no further painting, preservatives or stains, are weather- and rot-resistant and often easier to maintain. They also make use of a waste product that would otherwise be sent to landfill.

Go for greener grounds

Concrete may seem like a good, low-maintenance idea, but it's not. Aside from being ugly, it prevents seepage of rainwater into the ground, absorbs and reflects heat in summer and smothers all living things underneath it. Concrete also needs a substantial amount of energy to be produced. Instead, use ground cover, tan bark or paving that allows rainwater to seep through it.

Buy recycled

A huge number of landscaping products are the fruits of your recycling labour. Compost bins, retaining walls, garden furniture, garden boxes, edging, pots and even birdbaths made from recycled materials are all available. You can also use reclaimed bricks and pavers for paving. The fact that they're aged and a bit battered is the very thing that gives them their character.

> Mobile garbage bins or wheelie bins are made from 50% recycled plastic. It takes around 125 2-litre plastic milk bottles to make up the recycled component of a small 140-litre wheelie bin.

Use a greener grill

Even the humble barbecue can have a small but noticeable environmental impact. Charcoal-burning barbecues produce more air pollution and are less efficient than gas models. Go for gas instead. Electric barbecues are also available, though they typically don't reach the high temperatures that gas versions do.

Use light from the sun, day and night

Solar-powered garden lights with their own solar panels are ideal for providing outdoor lighting, as they do not require electrical connections. There is a wide range of solar garden lights available from hardware stores and some nurseries. Although they are relatively expensive to buy (starting at around $40 for a good-quality path light), they provide free, clean light.

LANDSCAPING TO SAVE ENERGY

The design and landscaping of the garden ultimately affects the house by reflecting or blocking sunlight, allowing cooling breezes, blocking cold winds and even insulating the house. Well-designed landscaping can make your home warmer in winter and cooler in summer, resulting in greater comfort and reduced heating and cooling bills.

Ideally, landscaping should be part of the design process when a house is built. Too often it is not even considered until the house is completed.

When planning your garden to improve household energy efficiency, you will need to know the following:

- prevailing seasonal winds and temperature variations
- direct sunlight and shade patterns for both summer and winter (the path of the sun is higher in the sky in summer), particularly in relation to the various rooms of the house
- existing trees and other vegetation, and whether they are deciduous
- the location and height of other buildings and fences on and off your property that may overshadow your house
- setbacks and building restrictions that may limit the garden architecture you put in place.

Controlling the sun

The sun provides welcome heat in winter, but overheats houses in summer. Consider the following tips for controlling the effects of the sun:

- Plant deciduous trees, or grow deciduous vines on a pergola on the north and west faces of your house. The bare branches will allow warmth and light in during winter and provide shade in the summer, reducing your need for additional heating and cooling. Alternatively, you can put up removable sail cloths or shade cloth to shade north- or west-facing windows in summer and take them down in winter.
- Windows are poorly insulated compared with walls, allowing the unwanted transfer of heat. A tree shading a window can reduce a room's temperature by up to 12°C in summer months. The highest priority is to shade windows, but it is also worth shading north- and west-facing rooftops and walls.
- The choice of shade trees is important. Deciduous trees allow a seasonal change in how much shade they provide. Trees with heavy foliage cast a dense shadow that is very effective at blocking heat. Trees with light foliage produce a more dappled shade, blocking some of the heat but allowing a certain level of light. Trees that grow to great heights may reach the point where they no longer shade windows but can effectively shade the roof.
- Horizontal shading, such as that provided by a pergola or a veranda, is ideal for north-facing windows. Pergolas or verandas can be built with shade provided by wooden slats, angled to run parallel with beams of winter sunlight, allowing it through, but blocking summer sun.
- Vertical shade is best for east and west walls and windows in summer. It protects the house from the intense summer sun at low angles. It can be provided by dense trees and shrubs or climbing vines trained over lattice screens.

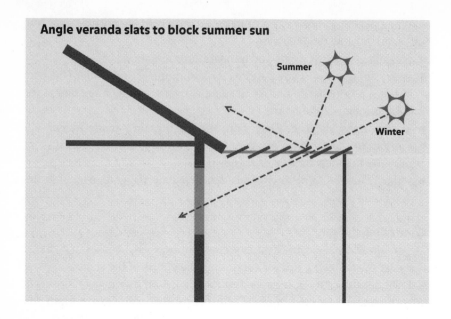

Angle veranda slats to block summer sun

- Strong sunlight can produce unwanted reflected heat and glare when reflected from light-coloured paving, walls, water or shiny surfaces. Vegetation absorbs more sunlight and re-radiates less heat than paved surfaces, and can lower the temperature near the ground by up to 6°C. If glare is a problem, consider replacing paving with ground-covering plants. Avoid paving near north-facing windows.
- If you want to retain paved areas that are producing nasty glare, shade the area by planting shade trees or installing shade sails.

Living insulation

Many people would consider ivy-covered walls and rooftop gardens a purely aesthetic choice. However, this kind of vegetation reduces the need for heating and cooling, and consequently reduces greenhouse gas emissions, by providing extra insulation and protection from wind. One Canadian study found that a 16-centimetre-thick blanket of plants can increase the level of insulation (R value) of a wall by as much as 30%.

Vegetated roofs or rooftop gardens and their substrates can also act as insulation in winter. In fact, there's a long history of this in Northern Europe— the traditional sod roofs of the Vikings are an example. But in Australian summers, vegetated roofs can make a huge difference. The leaves of vegetation provide shade, plus the evapotranspiration of water from plants has a natural cooling affect. A joint study between the University of Melbourne and CSIRO found that a vegetated roof can reduce the energy needs for a building's airconditioning by up to 48% and reduce energy needs for winter heating by up to 13%. Vegetated roofs also help to control pollution and damage caused by stormwater run-off by soaking up precipitation, particularly during heavy rainfall.

Controlling the wind

Like the sun, the movement of wind can work both for and against you, depending on the season and your local climate conditions. Consider the following tips for controlling the effects of local wind patterns:

- Plants, screens and fences can all help to block cold winds. However, solid windbreaks, such as fences and walls, can create turbulence. Over time the force of the wind and turbulence can weaken and damage walls and fences.
- Plant protective, semi-permeable, wind-breaking plants on the windward side of fences that are buffeted by strong winds. This will help to keep the fence upright and in good condition.
- Use semi-permeable windbreaks with 50–60% density. These are generally more effective than solid windbreaks.
- Hedges make effective windbreaks.
- Where possible, position windbreaks and other garden structures to encourage and allow the entry of cooling summer breezes. This may not be possible in locations where cooling breezes (such as sea breezes) and strong winds come from the same direction.
- Vegetation can help to cool hot summer winds. The transpiration of water from plants draws heat energy from the surroundings and so has a cooling effect—like a living airconditioner. Position plants and water features near shaded windows that capture summer breezes to harness this airconditioning effect.

PETS

Your four-legged friends might not always appreciate your efforts to green up their life. After all, stopping a cat from chasing native birds could be seen as spoiling its fun. However, other creatures, great and small, will benefit from your efforts.

Making your pets wildlife-friendly

Domestic pets, strays and feral animals claim the lives of thousands of native animals each year. It's not their fault; they're only doing what comes naturally. However, you can even up the odds and give the native animals a sporting chance.

- Feed your animals regularly (without overfeeding them!). If they're hungry, they're more likely to try to make a meal of other animals.

- Keep cats indoors at night.
- Put a bell on your cat's collar to give birds and small animals an audible warning.
- Have your cat or dog desexed. This will mean fewer unwanted litters. Desexed animals are also less inclined to wander and hunt.
- Never dump unwanted cats or kittens. Strays are forced to kill wildlife for food to survive.
- Make sure any pet birds you obtain come from reputable sources. The illegal capture and trade of exotic birds are threatening some species.

Eco-friendly pet products

You can be a green pet owner and a green consumer by choosing environmentally preferred pet products and avoiding those that are less than green:

- Buy biodegradable kitty litter. You can even get kitty litter that's made from wheat, corn or recycled newspapers.
- Buy more bicarb soda! Put a box out of reach in your dog's kennel or liberally sprinkle some between a pet basket and its lining. Baking soda absorbs smells.
- Remember the first line of attack when controlling fleas: stop the infestation from happening in the first place. Put some herbal flea deterrents on your pet's bed or basket, such as citronella, eucalyptus, pennyroyal or citrus peel oils. Also do regular flea combing so that you can spot the problem early, while there are just a few fleas to deal with.
- Buy herbal flea rinses instead of chemical rinses. Alternatively, look for flea collars that contain citronella and cedar oils or extracts from pyrethrum flowers, and add brewer's yeast and garlic to your pet's food to deter fleas.
- Particularly avoid flea rinses that contain the organophosphate chemicals chlorpyrifos or diazinon. These are highly toxic chemicals that often end up in waterways, harming aquatic life and sometimes the people and animals that have used them.
- As for your own food, try to prepare fresh food for your pets instead of using over-packaged processed food. It's better for the pet and better for the environment. You can even get some organic pet food products.
- If you do buy canned pet food, remember that the cans themselves are recyclable wherever normal food cans are collected.
- Keep in mind that a carnivorous pet will increase your household's carbon footprint, as the emissions from livestock are huge. You may want to carbon offset your pet.
- Consider buying a worm farm for dog poo. That way you can turn unwanted dog poo into worm poo, which is good for the garden. Have

one worm farm for food scraps and a separate one for dog poo. If you put the lot into one worm farm, the worms will go for the food scraps first, leaving the poo to rot and go smelly. Alternatively, flush dog waste down the toilet or toss it in the rubbish bin, using biodegradable bags.

- Don't put cat or dog poo in your compost.

→ **ASK TANYA**

I just wonder do you know how much do we add to the greenhouse effect by owning a pet, such as a dog?

Trung, Highgate Hill, Queensland

Good question! If we humans have a carbon footprint, then the furrier members of our households similarly have a carbon paw print. The size of your pet's greenhouse impact will depend on the type and size of animal, what you feed it, trips to the vet or obedience school, the products you buy for it and many other factors.

According to environmental retailer and consultancy Neco, the total emissions associated with the average medium-sized dog are 1.9 tonnes per year, and 400 kilograms of emissions per year for the average adult cat. But it can be much more. You can offset this through any one of a number of carbon offsetting programs.

Alarmingly, my local pet store now has a clothes rack. Yes, you too can buy designer duds for your chihuahua, just as Paris Hilton does (please read heavy sarcasm into this!). The dollar value of the pet product market is rapidly growing, particularly in the premium or luxury market. The landmark book Affluenza (C Hamilton and R Denniss, Allen & Unwin, 2005) identifies up-market pet products as a sign of our society's consumption binge. Pets need food, shelter, care and attention. They don't need pet cologne, diamond-set nine-carat gold nametags or treats that cost $100 per kilogram. Remember that every product you buy needed energy, water and material resources to make it and consequently has a greenhouse cost. Consumerism gone rampant is a big contributor to our increasing greenhouse emissions. With pets, as with everything, you can reduce your greenhouse impact by simply buying less useless stuff.

MAKE A COMMITMENT

This week I will...

☐ _____

☐ _____

☐ _____

This month I will...

☐ _____

☐ _____

☐ _____

When I get the opportunity I will...

☐ _____

☐ _____

☐ _____

Notes

_____

_____

_____

_____

_____

_____

12

Green building and renovating

I f you show some care for the planet when planning or choosing and adapting your house, you will be repaid with a comfortable, warm place in which to relax, with lower energy and water bills, and with a cleaner, healthier environment to live in.

Tens of thousands of new houses are built each year in addition to (or replacing some of) the millions of existing dwellings. If you're one of the thousands who are building a new house, then you have a unique chance to build a greener, more energy-efficient one that will take a lesser toll on the environment throughout its life. Good design features, such as the right orientation of the house on the block in relation to the sun, can make more of a difference to our energy use than the combined benefits of smaller energy-saving tips, such as putting lids on saucepans or turning off appliances on standby. Software has been developed that makes it easy for builders and architects to estimate how energy-efficient a new house is. There are also programs that train and accredit energy-smart and environmentally sensitive builders. It would be great to see eco-homes gradually become more common and the energy-efficient 'green house' become the normal way to build, rather than the exception.

This chapter is a snapshot of what to look for and what to ask your builder or architect to consider. You will need to seek expert advice because individual sites, settings and climate zones have particular considerations, and different local laws and council regulations will have some influence over what you can build in your area and how easy it is to get planning approvals. If you're building or renovating, you'll have to make some big decisions, but remember that there are many design features, sustainable materials and new technologies that are readily available. If you understand the issues, you will know what questions to ask.

WHAT MAKES A HOUSE GREEN?
There are new green houses that we tend not to notice because they don't always stand out and they don't fit the hippie stereotype. In fact, a green

house can look exactly like any other house. Green features such as solar panels on the roof or rainwater tanks out in the backyard are not all that noticeable.

The stereotypical hippie green house and the modern eco-home are both built around the same underlying principles of what makes a house greener. There are three key factors in greener buildings:

- **Energy efficiency** is about how well your house uses energy. It involves the appliances, heating, cooling, ventilation and lighting systems of your home. The design of the house and the use it makes of natural light and heat (passive solar heating), natural ventilation and insulation will also determine how much extra heating, cooling and lighting it will need. The term 'energy-efficient house' sounds quite practical and bland, so instead think of energy-efficient homes as comfortable, warm homes that are not expensive to run.
- **Water efficiency** involves water supplies (from water mains and rainwater), how water is used in the home and how wastewater is disposed of.
- **Material use** considers the sources of the materials used to build your house and whether they are sustainable, how they are made or processed before being used in the house, and their effect on how healthy and how efficient the house is to live in.

Looking at buildings from a different perspective, a green house doesn't necessarily make a green home. A green home is a sustainable house in action, fitted and lived in in an eco-wise way. The three factors that make a green home are:

- the house structure—incorporating aspects such as the design and materials used
- the fittings—including window coverings and appliances
- the occupants—the attitudes and habits of the people who live there.

The occupants are the human element of a green home. Just as a hammer can be a tool in the hands of one person and a weapon in the hands of another, a green house can still have a huge negative impact on the environment if the people who live there have wasteful habits and ignorant attitudes. Remember to adapt your lifestyle, as well as your house design.

With green houses, there are many shades of grey. A builder can achieve a house that goes a good way towards being better environmentally by putting in a bit of effort and sometimes a little extra expense up-front. Remember that energy-efficient houses pay back some of their costs through the energy saved over their life spans. Enthusiasts can go to more effort and make their houses even greener, but this can mean extra paperwork for planning approvals, more research and sometimes greater expense. Green house design is not a case of all or nothing. Aim to make your home as green as possible, within your means and the potential of the house or block of land. Think of the greening of your house as an ongoing project. Over time, replace outdated appliances with new, more efficient models, consider the environment when you're making renovations and, if possible, incorporate alternative water and energy technologies into your home.

Energy ratings for houses

The Energy Rating labels for appliances help you to choose appliances, such as refrigerators or washing machines that need less electricity to run. The fuel consumption label similarly helps you to choose cars that are more fuel efficient, meaning less money spent at the petrol station. Rating schemes allow you to compare the environmental performance of similar products.

There are now a number of rating tools (including software and sets of rules and/or mathematical formulae) that calculate a numerical rating for a new or existing house's environmental performance—a standard measure of the house's performance, sustainability or energy efficiency. Rating tools allow us to assess houses and set benchmarks and minimum performance levels. They make some green building credentials measurable, which in turn allows them to be regulated by government. Some rating tools are able to predict performance at the design stage. Others measure the actual performance of an existing house with people living in it.

If you are buying an existing house, ask if a rating assessment has been made of the house. This will become more common in the future, but it's worth asking about now. Rating tools for existing houses measure the performance of the house based on its design features and data from energy and water bills. They can give you a 'ballpark' idea of what future power and water bills will be like and how efficiently a house will run, allowing you to compare it with other houses.

BUILDING A NEW HOUSE

Building a new house is an opportunity to re-size your ecological footprint. Seemingly small considerations can make a big difference, so make sure you're well informed before making decisions and settling on a site and house plan.

Remember that the fundamental thing to get right is the design of the house in relation to the block of land it will be built on. Good green houses are the product of a perfect match between the design and the block. Many people fall in love with a particular house plan and then plonk it on any block that's the right size. Land can be hard to find, so it makes sense to choose the land first and then design the house to suit the land. There is a common misconception that the only blocks with good solar access are those that have a south frontage, allowing north-facing living areas at the back. In reality, a good architect can design a house with good access and north-facing living areas on just about any block. Some examples of good design appropriate to the block of land are included in the following diagram.

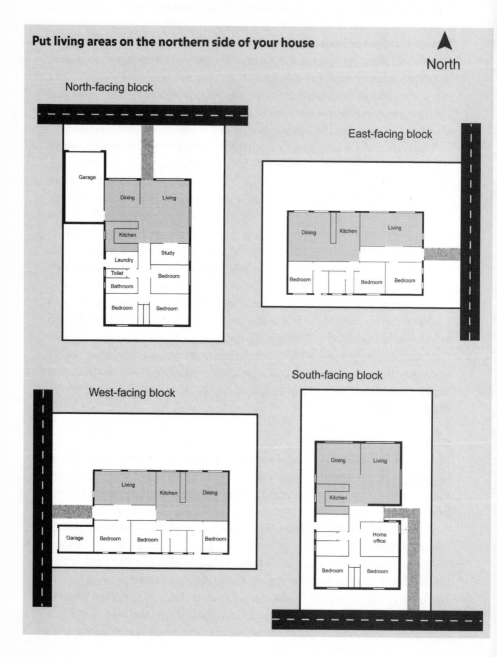

Put living areas on the northern side of your house

North

North-facing block

East-facing block

West-facing block

South-facing block

Environmental ratings for new houses

Energy ratings for houses are particularly important for new homes, as it's far easier to design a house for efficiency from the outset than to retrofit a poorly designed house down the track. Terms like 'six-star homes' relate to the thermal efficiency of a house—how well it keeps in heat in cold weather or keeps it out in summer. Most rating programs are related to the National House Energy Rating Scheme (NatHERS). NatHERS uses computer simulations to predict the thermal energy efficiency of a planned new home, providing the result as a number of stars, from one to ten. The more stars a house receives, the more comfortable it will be, with less need for heating and cooling. Most states and territories have implemented five-star efficiency as a mandatory requirement for new houses.

The average Australian home built in 1990 would have a one-star NatHERS rating. In 2009 the Council of Australian Governments (COAG) signed the *National Strategy on Energy Efficiency*. As part of this, the 2010 amendments to the Building Code of Australia (BCA) included the requirement for new homes to be built to six-star (or equivalent) efficiency. The details of how this is or will be implemented depend on the state or territory you live in, with some governments having a transition period to phase in the changes.

Governments are also looking beyond thermal star ratings to further regulate other aspects of home sustainability, such as lighting, hot water systems and the use of water-efficient fixtures. For example, all states and territories require new homes to have water-efficient toilets with a three-star water-efficiency rating or higher.

Don't be surprised if you hear mixed opinions on how good or bad five-star standards are. Organisations opposing these improved environmental standards argue that the additional features required to meet the standard make building costs higher for the consumer. However, the increased building costs are well and truly offset by reduced running costs. As the saying goes, 'the proof is in the pudding'. In Victoria, where there is a longer history of five-star homes, a survey of the experiences of 150 people living in five-star homes over a 12-month seasonal cycle found that nearly 90% would definitely recommend a five-star energy-rated home to others. Almost the same number, 89%, agreed that all new homes should be five-star rated; 89% said their homes were definitely warmer in winter, and 86% said that they were cooler in summer.

When you're consulting a developer or architect, find out the star rating for each design you are considering. Ask about various ways in which the rating can be improved. Don't just aim for six stars; see how high you can go. There are a number of different tools used by professionals to determine the star rating. Many focus on ratings that reflect the typical energy consumption of the house. However, some look instead at the greenhouse impact of the dwelling, which may be subtly different. Work is being done to diversify these tools so that they can also incorporate other environmental issues and benefits, such as improved water efficiency or indoor air quality differences.

Location

Choosing where you're going to live will immediately affect your ongoing greenhouse gas contribution. This has nothing to do with the house itself but with how far the house is from everywhere else that you need to go. The aim is to 'live locally' as much as possible.

Transport accounts for 34% of the direct greenhouse impact of the average household. The bulk of these greenhouse gases come from the exhaust pipes of cars, so look for a site near a train station or other public transport route so that you don't have to drive. If you live close enough, you could even walk or ride a bike to work in good weather.

Also look at the services available in the area. An established community with shops, schools and other facilities and services will reduce the amount of time you need to spend in your car.

Planning and design

Before you consider the specific design, think about what you need and want in your house. Consider the number of bedrooms you'll need now and in the future. Factor in storage space and, if applicable, a study or home office. Don't forget to plan for outdoor needs as well: allow space for things like a clothes line, compost bins, a rainwater tank, an outdoor entertaining

> **Home sweet home**
>
> In Australia there are over 8 million homes. Of these, 78% are separate houses, while around 9% are semidetached houses, terrace houses or townhouses, and the remaining 13% are flats, units, apartments and other dwellings.

area, a barbecue and garbage bins, with a path to the front nature strip. Now is also the ideal time to consider landscaping features that might help to improve the energy efficiency of your house, as outlined in chapter 11, pages 218–21.

Remember that you have to pay for every square metre of the building of your home. Each additional square metre will take more materials and energy to build. Once finished, each extra square metre is another to keep clean and another to heat. The frightening trend of building huge 'McMansions' and then having to heat them is part of the reason that Australia's household energy use is still increasing, despite our raised environmental awareness and our efforts to improve energy efficiency. Try to avoid designs that need a lot of hallways for access. Hallways are dead space. Instead, have areas of open-plan living, adequately closed off from the rest of the house to prevent draughts and heat loss. Maximise the amount of useful space, rather than wasting money and resources on hallways and unused rooms. Do your really need a separate home theatre room when there's already a TV, VCR and DVD player in the living room?

Orientation

Once you've chosen your land, it's time to decide how your dream home will sit on the block and which way the rooms will face. Harnessing the light and energy of the sun will save you a lot of money. The north faces of houses in Australia (and the southern hemisphere in general) get direct sunlight through most of the day. This sunlight can be used to help heat and light your home. Ideally, your block and the placement of the house on it should allow clear access to the north (often called solar access). It should not be overshadowed in winter by other buildings, tall trees or fences to the north. Solar access means your house will get the maximum free light and heat possible in the winter months. It can be restricted in summer by shade sails or deciduous plants.

The benefits of north-facing rooms and windows in the cooler parts of Australia are now widely known. Good use of a northerly aspect is free and will make your home brighter and more comfortable. It will also increase the eventual resale value of your house.

What goes where?

The location of each room also makes a difference. Consider the following tips:

- Locate rooms in which you spend a lot of time during the day on the northern side of the house. This includes living and family rooms.
- Put bedrooms to the south of the house. This side tends to be the coolest.
- The west of the house gets hot in the afternoon. Put service rooms such as the laundry, bathroom, garage and storage rooms on this side.
- 'Zone' rooms with similar uses and heating or cooling needs close together to make the distribution of heating and cooling easier to control and more energy-efficient. In a single-storey house, the three main zones are the living zone, the sleeping zone and the wet areas (kitchen, bathroom, laundry and toilet).
- Place your wet areas as close together as possible to reduce the length of pipes needed. This will also reduce energy costs by shortening the distance that hot water has to travel from the hot water system, losing heat along the way.

- If need be, place the bedrooms away from the front of the house to keep these sleeping areas away from traffic noise and pollution and to provide privacy.

Building materials

The choice of building materials is complicated and best made in consultation with your architect or builder. It requires expert knowledge and will depend upon the design of the house, the local climate and the condition of the land you're building on. From an environmental point of view, the following are the main considerations.

Material properties Different building materials have different properties that make them suited to particular climates and uses. For

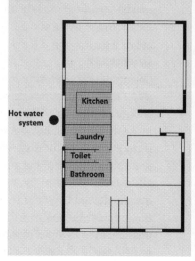

Place wet areas, together with the hot water system, near the kitchen

example, a humid climate will need materials that are moisture-resistant, while coastal areas will need materials that resist corrosion.

Durability As a general rule, the longer a building lasts, the less impact it has on the environment. Materials chosen should be durable, or easily replaceable should they have a shorter life span than the building as a whole. Often, higher-quality materials will cost more than cheaper alternatives, but they will save you money in the long term by needing less maintenance, not needing premature replacement and improving the house's resale value.

Renewable or recyclable resources Plant-based materials are renewable in that they can, in theory, be replaced by growing more, but our supply of resources such as metals is limited. It's important to make sure that, whenever possible, non-renewable materials can be easily recycled at the end of the building's life span.

Forest-friendly timbers Some natural building materials, particularly forest timber, are harvested from areas of natural vegetation. This destroys habitat, harms wildlife and has an impact on biodiversity. Always question the source of any timber that is used in your home and ask for alternatives to old-growth or rainforest timbers, such as recycled or plantation timber. For

further information on choosing the good wood, see page 234.

Energy efficiency Materials have varying degrees of energy efficiency too. For example, materials that have a high thermal mass (concrete, mud brick, rammed earth, stone and slate) are slow to heat and take a while to cool down. They help to regulate the temperature of your home by capturing and storing the sun's heat during the day and releasing it at night. High thermal mass materials can be used for floors in living areas that catch direct winter sunlight. The radiant energy from the sun comes through the windows during the day and is stored in the thermal mass of the floor. It is gradually released during the night, providing passive solar heating. In climates with cool winters, heavy building materials with high thermal mass can be used inside to help capture and store the sun's daytime warmth. For this reason, good architects will consider the size and placement of windows in conjunction with the use (or otherwise) of high thermal mass floor materials.

Indoor air quality Synthetic building materials, finishes and paints can all give off high levels of polluting gases, including volatile organic compounds (VOCs). These pollutants pose a significant health risk and can bring about symptoms such as headache, fatigue, respiratory problems, dizziness, and eye, nose, throat and skin irritation. Where possible, limit your use of carpets (particularly those made from synthetic fibres), synthetic adhesives, paints, varnishes (particularly those containing polyurethane) and other finishes, as well as particle board and MDF board. It can be very difficult to build a new house totally free from materials that give off these gases, so make sure that you ventilate the house regularly and thoroughly for the first year, particularly the first six months. Indoor air quality is covered in detail in chapter 2.

> **Ecological footprint**
>
> A person's ecological footprint is a measure of the impact that they have on the environment. It is an estimate of the area of land needed to produce the food and resources they use, and absorb the wastes they produce.

Waste and recycling Some materials and building methods generate a high amount of waste, while others can actually use recycled waste. Australians are among the highest producers of waste per capita in the world. Up to 40% of the waste going to Australian landfills comes from the building sector. Wherever possible, choose building or landscaping materials that are made from recycled aggregate or other recycled materials. Ask your builder about their waste minimisation practices, and consider incorporating waste minimisation and recycling clauses and incentives into the contract. You can also use building materials and features that have been salvaged from demolished buildings. Reclaimed bricks, for example, can add a lot of character to a home or garden and are sometimes cheaper than new materials.

Keep an eye out for unusual or interesting pieces that can bring interest and individuality to your home.

The house envelope

The barrier between the inside of your house and the outdoors is the house envelope. Windows, doors, walls, roofs and floors make up this envelope,

Timber: choosing the good wood

You can help to preserve our planet's precious rainforests and old-growth forests by refusing to use the timbers that are harvested from there. Instead, look for alternative and sustainable timber choices. Here's a quick guide to recommended timber choices.

Plantation timbers

Commonly available plantation species include:

- hoop pine (*Callitris collumelaris*)
- slash pine (*Pinus elliottii*)
- radiata pine (*Pinus radiata*)
- blue gum (*Eucalyptus globulus*).

Recycled timbers

- Commonly available species include blackbutt, jarrah, messmate, mountain ash, cedar, karri, ironbark, red gum, Victorian ash and many others. Note that 'recycled' timber does *not* include timber salvaged from land-clearing or forestry operations or dead native trees that might provide wildlife habitat.
- Avoid recycled timber that is chemically treated or impregnated with pesticides.

Agro-forestry timbers

Agro-forestry timbers are those established and managed as part of an agricultural enterprise, such as:

- blue gum
- cypress
- radiata pine.

Timber alternatives

Alternatives to timber include:

- bamboo—commonly used in flooring and furniture
- recycled plastic 'lumber'—often used in decking
- wheat-based fibreboards.

There is also a small number of timber suppliers that are certified in line with the Forest Stewardship Council (FSC) standards. FSC is an international non-profit organisation working towards the environmental, social and economic management of the world's forests. FSC in Australia is in its early days, but hopefully we will see it become more prominent Down Under. For more information, visit www.fscaustralia.org.

Be wary of other timber 'eco-labels', as many are self-evaluating voluntary initiatives run by the forestry industry and/or do not require independent auditing.

protecting you from the elements, temperature extremes and some of the forces of nature. The ideal envelope is well insulated and airtight, has windows chosen to balance the need for natural light with the need to control heat loss, and allows controllable ventilation to ensure healthy indoor air.

Insulation

Insulation is essential to keep your home at a comfortable temperature throughout all seasons. Depending on the design of the house, it can be used to line the roof, ceiling, attic, external walls, basement, crawl space and floor. You may also wish to consider acoustic insulation to block out noise pollution, particularly if you have a music room in the house (perhaps with a drum kit) or if you live near a busy road or railway line. Soft furnishings, such as rugs, curtains and wall hangings, can also provide some insulation.

Insulation comes in two types, bulk and reflective. Bulk insulation uses porous materials that trap air in small pockets to reduce the flow of heat. Just about any material can be used to make insulation, as long as it can trap air. Many insulation products are made from recycled materials, making them even greener. Reflective insulation, as the name suggests, works by reflecting some of the radiation that falls on it. Both types can be used in combination to maximise the insulating effect.

Some people try to skimp on insulation, especially when the building budget is tight. This is a false saving as you pay more in heating and cooling bills over the time you live in the house, not to mention the discomfort of being too hot or too cold. Particularly consider insulation for external walls and floors when you're building because it is harder to install in existing constructions than ceiling and roof insulation.

Heat-leaking houses

Whenever you turn on an airconditioner or heater in a poorly insulated home, your money disappears into thin air.

Insulation is made and sold in imperial measurement (R) values, a measure of the insulation's resistance to heat flow. The higher the R value, the better insulation the material gives. The Building Code of Australia now requires certain levels of insulation in different climate zones in Australia. They can be found online at www.yourhome.gov.au under 'Insulation' in the 'Passive design' section of the *Technical Manual*.

If your house is built on a concrete slab, the edge of the slab should also be insulated, as a single square metre of exposed slab can lose as much heat as several square metres of uninsulated wall. This is particularly important in homes with in-slab heating systems, which can lose even more heat and waste huge amounts of energy.

Insulation should be considered as an integral part of the whole house envelope. Considering the amount of heat loss from large, single-pane windows, it would be pointless to add a lot of extra wall insulation to a house that has this type of windows.

Windows and shading

Windows are a mixed blessing. On the one hand they allow light and warmth in during winter. They also let fresh air in and pollutants out. On the other hand, they permit undesirable heat transfer. In fact, windows can be the cause of around 10–20% of heat loss in winter and around 25–35% of heat gain in summer.

Windows can be made more energy-efficient through the addition of an extra pane of glass, known as double-glazing, which stops some of the heat transfer. They can also be further insulated with thick curtains and well-fitting pelmets.

Recent years have seen technological developments with windows that further improve their performance over the old, cold conventional single panes. One development is the low-emissivity (low-e) coating—a thin, transparent layer of metal oxide over the glass, which allows visible light through but reflects back infrared heat radiation. This helps the window to keep heat out in summer and reflect heat back into the house in winter. It also has the added benefit of reducing incoming ultraviolet light and so protects furniture and drapes from fading. The space between panes in multiple-glazing can also be filled with argon gas, an inert gas that provides more insulation than air. Gas filling in combination with a low-e coating is particularly effective at improving the insulating level of windows.

Size and placement of windows and how they are shaded will depend on their orientation and the room they're placed in. The materials the window and its frame are made of, and how well they are fitted, will also influence how well the windows perform.

Tips for windows

- If there is good solar access, use large double-glazed windows for north-facing living areas.
- Keep windows on the south face of the house a small to medium size.
- Put small windows on the east and west faces of the house, with good summer shading. The east gets direct sunlight in the morning hours, while the west gets hot, direct sun in the afternoon.
- When selecting windows, look for the Window Energy Rating Scheme label, which is like the Energy Rating label for whitegoods. For window

recommendations for your climate area (within Australia), go to the scheme's website at www.wers.net.

- Position windows that can open and close in places that allow for cross-ventilation in living areas.
- Use double- or even triple-glazing for windows that are high up and hard to get at, such as skylights and clerestory windows. These windows may be hard to put curtains on or may not be intended to have curtains for aesthetic reasons.
- For the framing material, metal frames (usually aluminium) without a thermal break allow heat loss and gain. If you choose metal frames, make sure that they have some form of insulating treatment. Wooden frames insulate well but need more maintenance than aluminium. PVC frames are low maintenance and insulate well, but are made from non-renewable resources and contribute to pollution in their production. PVC in general should be avoided.
- Always seal gaps between the window frame and the wall to reduce air leaks, draughts and heat loss.

→ ASK TANYA

We have purchased a 1950s house in Adelaide with a north-facing backyard. We plan to do an eco-renovation and want to include a polished concrete floor in the lounge area (the northern aspect of the house). The plan is it will capture the northern winter sun and passively heat the area. We are thinking about floor to ceiling windows on the north-facing wall, so my question is: would double glazed windows reduce the amount of sunshine that could enter to heat the concrete floor in winter? We thought double-glazing would help with insulation, but want to maximise the warmth from the sun in winter.

Jane, North Plympton, SA

First, you're doing a great thing environmentally by choosing to eco-renovate an existing house rather than bulldozing it and building a flash new one from scratch.

It's a great idea to capture the free heat from the sun, and having a north-facing yard gives you a good opportunity to do so, provided it's not overshadowed by large trees or neighbouring buildings. Assuming the glass isn't tinted or reflective, double-glazing won't noticeably reduce the amount of radiant heat–producing sunlight pouring through your windows.

The level of insulation is given as an 'R value' (the higher the number, the greater the insulating effect). A single pane of glazing has an R value of 0.14. Double-glazing brings the window area's R value up to 0.31. Curtains will also provide insulation, as well as privacy, potentially bringing the R value of the covered, double-glazed window up to 0.45 for conventional curtains with pelmets. Window frame materials can also have an effect.

Capturing the sunlight will help to keep you warm in winter, but you won't like it in summer unless you factor adequate summer shading into your plans. Having spent several years in Adelaide, I know how hot it can get. External shading is the most effective way to keep out summer heat. Options include an eaves overhang, sail or shade cloths that can be removed in cooler months, awnings and pergolas.

- Fit windows that can be opened with flywire screens to keep insects out and to remove the temptation to use bug spray.
- Use window coverings that will make your house more energy-efficient. For more information, see the window-dressing tips in chapter 9, page 130.

Tips for shading

- Make sure that your roof has eaves that overhang the north face of the house. During winter, the path of the sun is low enough in the sky to allow light and radiant heat into the room. However, in the summer, when the sun is higher, the eaves will block the sun and help to shade the room.
- External blinds can effectively prevent much of the summer heat coming through the windows. Particularly consider them if you don't have much shade on the west side of the house.
- Deciduous plants can provide shade in the summer and lose their leaves and allow the light through in the winter. Consider planting these to the north of your house or near windows on the east and west faces of the house. Removable shade sails or shade cloths can similarly provide shade in the summer and be taken down in the winter.
- If you live in a hot climate, where northerly solar access is undesirable, consider building your house with a veranda encircling it.

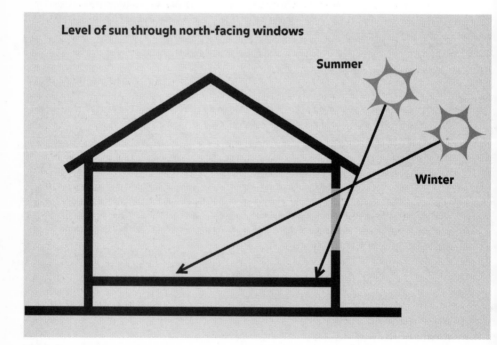

Fitting the interior

Flooring

Floors can be covered with a whole range of materials. When choosing a floor covering you need to consider the durability of the flooring material, how easy it is to clean and maintain, how it affects your health and that of your family, how sustainable the material is and the special flooring needs of particular rooms. Some materials are better suited to some rooms than others.

Floorboards are easier to clean than carpet and discourage dust mites. They can also be better for the quality of indoor air compared with gas-emitting or dust mite–ridden carpet, provided they're not finished with an off-gassing synthetic varnish. If you're looking at timber floorboards, consider the option of using recycled or plantation-grown timber.

Carpets are often preferred in bedrooms because they're softer on bare feet, absorb noise and help to keep out draughts. However, they can provide the right conditions for dust mites to grow in, so carpets shouldn't be used if you have family members with respiratory problems. Synthetic carpets, like other synthetic materials, produce gas pollutants, and even pure wool carpet often has a synthetic underlay or chemical treatment that can contribute to poor indoor air quality. Consider washable rugs as an alternative.

Natural fibres including sisal, coir and jute make sustainable and low-allergy floor coverings.

Resilient flooring is soft flooring that feels rubbery and comes as tiles or roles of sheet flooring. It is generally made from vinyl, rubber, cork or linoleum (lino). Resilient flooring comes in a wide range of designs, finishes and prices. It is often used because it is fairly easy to install and maintain and is relatively cheap to replace.

Eco-carpet

If you're looking for carpet for a school, government or commercial building, have a look at the award-winning InterfaceFLOR systems. InterfaceFLOR has a unique and sustainable approach to providing flooring systems, providing carpets made from recycled materials, suited to high-traffic areas, and offering the option of replacing and recycling just the worn-out carpet tiles (leaving in place any carpet tiles in good condition in low-traffic areas). Visit www.interfaceflor.com.au for more information.

Ecospecifier and GreenTag

Want to find some greener building and interior materials and don't know where to start looking? Look up the Ecospecifier. This is a fully searchable online database of environmentally preferred materials and where to get them. It includes listings for recycled plastic and hemp textiles, bamboo flooring and a huge range of other housing materials. Each listing gives an overview of the environmental benefits and where the materials can be used. Ecospecifier also runs the GreenTag environmental certification program for building materials and products. Go to www.ecospecifier.com.au.

However, it can be thin, can be easily damaged and can show irregularities in the surface over which it is laid. Care must be taken when choosing the adhesives that stick the flooring to the subfloor as some adhesives off-gas high levels of VOCs. Look for water-based adhesives where possible. Resilient flooring products made from the natural materials cork, rubber and lino are renewable and more sustainable. Vinyl and other PVC products are generally polluting in their manufacture.

Cork is a kind of resilient flooring with a wonderful environmental story behind it. Cork is a natural, sustainable material harvested every nine years from the bark of the cork oak tree. It grows without the need for chemicals, fertilisers or irrigation. Cork is also non-polluting, biodegradable and recyclable; what more could you ask for? Cork tiles help to insulate the floor and are soft to walk on. If you decide to use cork flooring, avoid tiles that are PVC-finished. Consider buying them unvarnished and finish them yourself with a water-based varnish or beeswax polish. The fact that they come in tiles makes it easy to replace any sections that become damaged, so make sure you buy a few extra.

Bamboo is another natural floor covering with some green credentials. It is an incredibly fast-growing plant, making it a renewable and quick-to-produce green alternative to timber floorboards. Bamboo also replaces 30% of its biomass every year, and bamboo forests reportedly sequester 2 tonnes of carbon dioxide per hectare, more than five times the amount sequestered by an equivalent group of trees. Bamboo floorboards function similarly to hardwood floorboards. Their key benefits are that they are moisture-resistant, very strong, durable, don't buckle or twist, are free of knots or large flaws, and are naturally a light blond colour that can be stained to produce the effect of darker woods if desired. Bamboo flooring is cost-effective and readily available in tongue and groove strips. It can be bought both unfinished and polyurethane-finished.

Slate, stone, ceramic and terracotta tiles, and exposed concrete are popular flooring materials, particularly in the wet areas of the house. Because they are made from mineral resources mined from the earth's crust, they are not renewable, but some of the raw ingredients, like the sand in concrete, are in abundant supply. Others, such as some kinds of slate, granite and marble, are more rare. The main environmental impacts from these materials are the damage done to local ecosystems when they're extracted, the energy and water used and waste produced in their processing and, being heavy materials, the fuel and greenhouse costs of their transportation. Materials that have to be fired in a kiln, such as ceramic tiles, also have a high energy cost.

For these reasons, stone and ceramic flooring products are a good choice environmentally when they're recycled or salvaged stocks, sourced from a local supply. Other benefits are that they are generally easy to clean without the need for harsh cleaning agents and they also provide better indoor air quality than carpets, as long as they haven't been finished with fume-emitting varnishes or sealants. They don't harbour dust mites.

You can also get various kinds of aggregate or 'engineered stone', a mixture of mineral materials (such as quartz, marble or granite) and binding resins. They tend to be long wearing and are often made in non-porous varieties. One environmental benefit of engineered stone is that it can be made with some recycled or salvaged content, such as waste marble, recycled coloured glass, and even oyster and mussel shells.

Green fittings

Nowadays there's a lot more to a house than four walls and a roof. Most houses also come with heating and cooling systems, a hot water system, cooking appliances, toilets, baths and showers, and a range of tap fittings. All of these can be chosen to reduce your ongoing use of energy and water. Here's a snapshot of what to consider:

- **Heating systems** Particularly consider whether or not you would like an in-floor heating system, such as hydronic or electric slab heating, or a ground-source heat pump. These generally cannot be fitted later without considerable cost and work. For more information about choosing a heating system, see chapter 9, pages 131–8.
- **Cooling systems** For more information about choosing a cooling system, see chapter 9, pages 142–4.
- **Hot water systems** Locate the hot water system as close as you can to the kitchen to reduce the amount of cold water sitting in lengths of pipe that gets wasted while waiting for the hot water to come through. Hot water will also lose more heat travelling through a greater length of pipe on its way to the place of use. If you're considering a solar hot water system, make sure you factor this into the roof and plumbing design for the house. For more information on choosing a hot water service, see chapter 9, pages 144–8.
- **Water-using devices** For information on water-efficient fixtures and the water-efficiency rating program, see pages 182–3.
- **Cooking appliances** For more information on choosing cooking appliances, see chapter 9, pages 155–6.

How hard is it to build green?

Fortunately, the government now wants us to build more energy-efficient houses so that we can reduce our greenhouse emissions and attempt to meet Kyoto Protocol targets and government commitments. Six-star efficiency is becoming the norm for home builders, and there are display homes for designs that achieve eight stars or more. More builders and developers are appreciating the competitive edge that greener, more comfortable, more energy-efficient homes give them, and are developing project homes with higher energy ratings.

Look out for display homes and house plans that are designed with energy or the environment in mind. Don't assume that the designs on offer are set in concrete and can't be changed. You'll find that developers are often willing to modify standard plans, for instance by changing the placement of windows or by increasing the size of eaves at little or no additional cost. You can also pay a little extra to incorporate other environmental features, such as additional insulation or an upgrade of the hot water service to a higher-efficiency or solar system. Don't be afraid to ask. Even if they say no, your interest in environmental concerns will often be registered and they may consider designing greener alternatives in the future. Change often begins with enlightened people asking for it.

BUYING AN EXISTING HOME

When you're looking at existing houses to buy, remember that renovations can turn a moderate house into a greener house, but all renovations come at a cost, both in time and money. Don't forget that it's the existing house that you'll take possession of and possibly move into, not the house in your imagination.

Here's a green checklist of what to look for in an existing house.

Green checklist for house hunting

Design

- Does the design suit the climate? For example, if it's a cooler area, is there good solar access, or if it's in an area that gets hot summers, is the house adequately shaded with eaves or verandas?
- Are the living areas on the north side?
- Are the bedrooms on the south side?
- Are the service rooms on the west or south sides?
- Does the placement of windows allow for quick cross-ventilation?

Energy

- How much natural light is there in the living areas?
- Is the house draughty? Are there cracks or gaps in the walls or floors?
- What is the current heating system? How efficient is it?
- What other energy-using systems are in place?
- If it's an older home, have energy-efficient improvements or renovations been made?
- Overall, how energy-efficient is the house and how much electricity and gas does it typically use? You can ask to see past energy bills to get an indication of what it costs to run, taking into account the previous number of occupants and their particular living habits and appliances.
- Is the house well zoned to make heating and cooling easier?
- How well is the house insulated? Can more insulation be added if needed?
- Are the windows single-pane or double-glazed?

Water

- What condition are the plumbing fixtures and pipes in?
- Are the plumbing fixtures water-efficient?
- Is there a rainwater tank or a convenient place to put one?
- Is the garden drought tolerant? Does it suit the local climate, or will it need watering?

Interior

- What are the floors covered with? Are the coverings in good condition?
- What appliances are included in the goods and chattels?
- Have any toxic materials been used inside, such as lead-containing paint?

If you come up with satisfactory answers to most or all of these questions for a house you're considering, buy it. If not, then weigh up what is lacking against your ability and budget to change it. Sometimes it can be better to buy a smaller house that's on the right track and in the right location and add an extension later. Always get professional advice and an estimate of the costs of changes by having a professional builder or architect look at the house before you buy.

GREEN RENOVATIONS

Renovation is a way to bring new life to an old home. It's also a way to stay in a loved home as a family outgrows it. Green renovations can improve the comfort and efficiency of your home. This might include incorporating alternative technology, such as a new solar hot water system. Many of the green ideas for building a new home that we've seen in the first part of the chapter can be applied to an existing house. The added dimension is that there is an existing house with solid structures to change, move or demolish, rather than a blank canvas. While your aims may include helping the environment, this process can instead harm it. A few simple green renovating tips will help you to change your house into your dream home without producing an unhealthy home environment or needlessly harming the planet.

Bringing in an expert

For most people, a house is the biggest purchase they will make in their lives and is their greatest asset. As a home owner, its up to you to make sure that your house is safe, in good condition, healthy to live in and not too expensive to run. Your house may contain an accident waiting to happen or a heater guzzling more energy than necessary—all without you knowing it. Just as a doctor or naturopath can give your body a check-up and uncover any undiagnosed health problems, a professional home inspector, home sustainability assessor, builder or architect can give the house you live in or one you're considering buying a building check-up. It is your responsibility to know or find out what state your house is in.

Archicentre architects can also provide professional, independent services to home buyers, home builders and home renovators. Archicentre is the building advisory service of the Royal Australian Institute of Architects.

Archicentre offers pre-purchase home inspections, in which an architect carefully inspects the house using a comprehensive 200-point checklist. The report identifies any components that aren't performing properly, safely or efficiently, any items that need replacing or to be repaired and any health and safety risks. You can also use the report to help in negotiating a purchase price for the house.

Archicentre pre-purchase house inspections generally cost around $500 for an average three-bedroom home (more or less, depending on the size of the house) and take around an hour. They need to be booked around two working days in advance. In some areas, Archicentre also offers home safety inspections. Visit www.archicentre.com.au for more information.

You can also get a home sustainability assessment to help you identify where you can make easy improvements to your home's energy and water efficiency. Choose an assessor accredited by the Association of Building Sustainability Assessors. Their website at www.absa.net.au has an online database to help people find an accredited assessor.

Green renovating tips

Sanding should be done with a sander with a vacuum attachment. Wear a mask to protect your lungs. Take particular care if you're sanding wood that has been treated, as the fine wood dust particles will also contain potentially harmful chemicals.

Paint stripping should be done with water-based, low-allergy paint strippers. Avoid conventional paint strippers, particularly those that contain dichloromethane (DCM), which is believed to be carcinogenic.

Paint containing lead may be found in houses built before 1970. Lead is a highly toxic element. Exposure to even the tiniest amounts of lead in dust or debris from renovations can cause health problems. Hardware stores sell home kits that help to test for the presence of lead. If your home does have paint containing lead, get professional advice before removing or disturbing it. If you do plan a renovation that will disturb it, then move out of the house until the renovation is finished.

Wallpapering can retain moisture and lead to fungal growth. Mix borax into wallpaper paste to prevent this instead of using a fungicide.

Painting can be made less wasteful by getting your estimates right. Look for paints certified by the Good Environmental Choice program. Alternatively, choose plant-based paints and finishes as they're made from sustainable materials rather than petrochemicals. They also produce far fewer indoor air-polluting fumes. Plant-based paints avoid using volatile petrochemical solvents and consequently can take longer to dry. Allow for this when planning your renovations.

Don't paint. Consider opting for natural, unpainted finishes where possible. Some porous materials such as cork and wood can be adequately sealed and preserved with natural oils or beeswax.

Recycle. Carefully remove any slate or ceramic tiles, aiming to keep as many intact as possible. Recycled tiles can often be sold second-hand or reused in landscaping or other areas of the house. If you feel like getting crafty, keep broken tiles for making mosaics.

Clean up safely. Never pour unused paint, solvents or water from washing brushes into stormwater drains as this can seriously contaminate groundwater and nearby streams, wetlands, lakes, rivers and oceans. Unwanted paint and solvents should be disposed of in an environmentally responsible manner. Some states, regions and local councils have hazardous and liquid waste collections, so contact your local council as the first port of call for specific advice. Unused paint can be disposed of by brushing leftover paint onto sheets of old newspaper. Brush one layer, then put the next layer directly on top and brush that one too, and so on. Let the paper dry and throw it out with your normal rubbish. In some areas, dry steel paint cans can be collected for recycling with other steel food cans.

Always remember to clean up thoroughly after any renovating, making sure that you vacuum and clean up every skerrick of dust or spilt paint. Over the following days, vacuum frequently and air the house to make sure that you capture any solid particles that have escaped you earlier and to allow any potentially harmful fumes to dissipate. That way you can enjoy your greener, new-look home without risk to your health.

→ ASK TANYA

We're about to do some renovating. What should I look for in house paints?

Peter, Doncaster, Victoria

Paints and architectural coatings have the potential to pollute both your home and the broader environment. Oil-based paints contain solvents, such as turpentine (or 'turps') and other volatile organic compounds (VOCs), and require them for cleaning up brushes and paint rollers. These solvents are flammable and can cause respiratory irritation, headaches and dizziness.

Look for water-based low- or no-VOC paints, such as those certified by the Good Environmental Choice Australia program, which sets a tough standard for VOC levels. This is particularly important with interior paints, as they off-gas into an enclosed space that you will ultimately be living and sleeping in.

Your choice of colour can also make a difference. Lighter colours reflect light and heat, while darker colours absorb it. Using very dark interior colours can make a room hotter in summer and can increase the need for artificial lighting.

MAKE A COMMITMENT

This week I will...

☐ _____

☐ _____

☐ _____

This month I will...

☐ _____

☐ _____

☐ _____

When I get the opportunity I will...

☐ _____

☐ _____

☐ _____

Notes

_____

_____

_____

_____

_____

_____

13

Getting around

There's no nice way to put it. Cars are bad for the environment. Ideally, we should all stop using our cars or, at the very least, use them much less. But with Australia's sprawling suburbs and growing population, this is a major challenge.

Drive-by polluting

Most of us wouldn't dream of throwing rubbish out of the living-room window into the front garden. Yet it appears that Australian motorists have no problem with trashing our shared backyard. In September 2010, Keep Australia Beautiful released an update of the *National Litter Index* report. Researchers found that motorists are among Australia's worst litterbugs, throwing a substantial portion of the nation's litter on highways and in car parks. Common litter items include cigarette butts, bottle tops, confectionery wrappers, tissues, snack food bags and can pull rings.

Do the right thing when you're on the road. Either avoid eating in your car (particularly if you have small children) or keep a small rubbish bin or bag in the car and empty it at the end of the trip into a proper bin.

When most people think about cars and the environment, they think about pollution. Car exhaust contains combustion gases and fine particle emissions—in the past, it also contained lead. All of these gases contribute to air pollution by reacting with sunlight to form photochemical smog, which is bad for our health and can trigger asthma attacks. Smog and air pollution are particularly concentrated in urban areas. All of us have to breathe, so it's no wonder that urban air pollution is such a cause for concern. Urban air quality directly affects the health of a city's residents. The World Health Organization estimates that air pollution causes approximately 2 million premature deaths around the world each year.

The other problem is the greenhouse effect. Greenhouse gases such as carbon dioxide, methane and nitrous oxide float up into the atmosphere where they act like a nice cosy blanket for the planet, preventing heat from escaping the atmosphere. This leads to global warming, which changes the climate. One of the scariest things about the greenhouse effect is that it doesn't respect national borders and can't be confined to a region. The effects of a few industrialised countries performing poorly on greenhouse issues are felt by the whole world. Australia and the USA are among the highest producers of greenhouse gases per capita, making us very unpopular with our Pacific neighbours, particularly those in the vulnerable low-lying islands.

Transportation accounts for 34% of the greenhouse gases produced by the average household. Though some of this is from public transport and air travel, a high proportion is from private car use, which we can modify for the benefit of the planet.

When you're buying a car, it's important to consider all aspects of a car and how they affect the environment. From an environmental perspective, your choice of car is more important than the way you drive it and maintain it. Once you buy a car, you're locked into that particular car's fuel consumption and exhaust patterns for the entire time you use it. For more information on buying a new car, see page 255.

Cars have a life cycle, each stage of which has different environmental implications. For example, the production of the car uses resources to make the materials that make up the car and its parts, and energy is required to mine or produce these materials, to transport them to the factory, in the manufacturing process itself and, finally, to transport the finished product to the buyer.

The car owner has control of the car during its use. This is where you, as a consumer and a car owner, can make a huge difference. One MIT study found that 80–90% of a car's total energy use occurs in its 'operational stage', compared with the 7–12% required for materials production and manufacture. So you're in the driver's seat when it comes to minimising energy use. While your car is on the road, its exhaust will contribute to pollution and to the greenhouse effect. Caring for and cleaning the car will use water, detergents and other chemicals. These environmental impacts can all be reduced.

Finally, once the car is no longer wanted for transportation and has reached the end of its life, it is dismantled. Reusable parts are removed for use by mechanics; steel and other marketable materials are collected for recycling, and the remainder is disposed of.

To make a difference, you need to look at how you can soften the environmental impact of your car, as well as considering the latest green machines on the market, alternatives to car transport and the future for car transport and fuel.

SAVING FUEL

The amount of fuel you use relates directly to the amount of greenhouse gas your car produces, regardless of which car you're driving or which fuel it's using. In short, no matter what car you currently have, it will help the environment to reduce your fuel use. The bonus is that it will also save you money.

Plan your driving

Save fuel and time by doing a number of tasks in one trip rather than making a separate trip for each task. You can also turn local trips (for example, to get milk when it unexpectedly runs out) into healthy exercise by walking, riding or rollerblading rather than driving. Try to avoid driving during peak hour, as the stop-start driving that you do in heavy traffic is very inefficient. Sitting in peak-hour traffic can be stressful and unhealthy, particularly if you're sitting behind a large truck, taking in its dirty diesel exhaust.

Drive smoothly

Don't push your car too hard. In manual cars, move up through the gears with comfortable acceleration. Don't drive in a higher or lower gear than necessary, as this wastes fuel. In both manual and automatic cars, try to maintain a steady speed and avoid rapid acceleration and deceleration, which guzzles petrol. A European study showed that aggressive driving—fast acceleration from traffic lights and hard braking—reduced travel time by only 4%. This is equivalent to only 2.5 minutes out of a 60-minute trip, resulting in 57.5 minutes of stressful and somewhat dangerous driving.

Don't speed

Slower driving is also greener driving. High speeds use a lot of extra fuel. For example, according to the Australian Greenhouse Office, at 110 kilometres per hour your car can use up to 25% more fuel than it would at 90 kilometres per hour.

Don't just sit there ... stop idling!

Many people needlessly waste petrol and produce a lot of pollution by leaving their engine running to warm up the car in winter. Once your car is running, the best way to warm it up is to drive it. Late-model cars with computer-controlled, fuel-injected engines don't need more than 30 seconds of idling before driving in cold weather.

Keep the tyres pumped up

Slightly flat tyres have increased resistance against the road, creating a slight drag on the car. This means that you need more fuel to go at the same speed. It also reduces the life of the tyre. Check regularly that your tyres are at the recommended pressure, particularly in winter, when cold temperatures decrease the air pressure. A single tyre under-inflated by 6 pounds per square inch (psi) can increase fuel consumption by 3% and take 10,000 kilometres off the tyre's life.

Turn off the engine when stationary
If you're in a traffic jam and not moving, turn off the engine. Similarly, turn it off when you're stopping for more than ten seconds. For example, when pulling over to collect someone, don't leave the engine running while waiting for him or her to show up.

Have your car serviced regularly
As with all machinery, cars run more efficiently when they are in prime working condition, whereas out-of-tune cars burn more fuel and produce more polluting emissions. They're like us that way: we work better and get more done when we're healthy. Make sure that you get your car serviced and tuned at the recommended intervals.

Don't overfill with petrol
Spilling fuel by overfilling at the pump and evaporation from a poorly sealed tank can both waste petrol. Fill up to the first click and make sure your car's fuel cap fits properly.

Lighten the load
The more weight your car carries, the more fuel required to move it. Remove any unnecessary weight from your car. For example, when you get home after a road trip, unpack your car straightaway. If you live in the snowfields of Australia or New Zealand (or elsewhere), clear any snow piled on top of your car or snow that has built up under the bumper or in the wheel wells. Snow adds weight and increases the aerodynamic drag on your car.

Reduce the need for airconditioning and heating
Using the airconditioner, particularly on hot days, can increase fuel consumption by around 5–10%. On hot days, when the interiors of cars left in the sun reach roasting temperatures, drive for the first few minutes of the trip without the airconditioner on, and with the windows down to let the hot air inside the car escape. Then turn on the airconditioner. Switch over to the economy mode once the car is cool. Park in the shade or under cover wherever possible, and put sun shades up in your car windows when you leave your car in the sun.

Conversely, leave the car in the sun in winter to heat up the interior in the same way a greenhouse heats up. You can also reduce the need to use the heater in winter by dressing appropriately—if it's winter, don't expect to get around in a T-shirt.

Use the boot, not the roof

If there's space in the back or boot of the car, use this instead of putting loads onto roof racks. Items on a roof rack will increase wind resistance, reducing the aerodynamic efficiency of your car.

Green-clean your machine

It's easy to justify cleaning your car. Done properly it will help to keep the paintwork in good condition, prevent rust and ensure a better resale value, should you eventually sell it. When you clean your car, make sure that you're using water wisely and that you're not releasing pollutants into the waterways and the environment.

- Waterless car cleaning products are available, though there is very little information available about exactly how much water they need to be manufactured. Favour those that are certified by the Good Environmental Choice Australia program.
- Wash your car on the lawn to stop water and detergents from flowing into stormwater drains and eventually into the nearest stream, lake or the sea.
- Use buckets instead of a hose for washing and rinsing. You can also use rainwater or greywater. In some states, car washing with a hose is banned. If caught, you could be dealt a hefty fine. In some areas, hoses with trigger nozzles may be allowed for car washing. If in doubt, check with your local water authority.
- Another alternative is commercial car washing stations. Though they use some water and detergents, the use of small nozzles and pressurised hoses reduces the water consumption, and the management of their wastewater is regulated. Some recycle their reclaimed wastewater. The Australian Carwash Association runs a car wash water-rating scheme for its members, which is independently audited. If you choose to use a commercial car wash service or to clean your car in a self-service car wash bay, look for a business that achieves a five-star rating under the scheme. For more information, see www.waterratingscheme.net.au.

RECYCLING

Used motor oil

One litre of dumped used motor oil can contaminate a million litres of fresh water. If you change the motor oil yourself, never pour it into the gutter or down a sink or drain, and never dump it. If not properly disposed of, it can lead to the contamination of soil, air, groundwater and other waterways. Used oil or 'sump oil' is now being recycled in Australia, and many councils have drop-off centres that take unwanted motor oil and other liquid wastes. The government-backed 'Product Stewardship for

Oil' program has a directory of oil recyclers and collectors online at www. oilrecycling.gov.au. Your local council can also give you information about local used-oil collections.

Retiring old cars

The old stereotype of the hippie greenie driving a beat-up old seventies kombivan is actually far from green. While the hippie may be doing many things to reduce energy use and improve air quality, the old van will be working against these efforts. One of the best things you can do for the environment is to retire any old, smoky bombs you may have. With antique and collectors' cars, you can have certain modifications made to make the exhaust less polluting.

An ageing population—of cars

Despite record new car sales, the average age of cars in Australia is among the highest in the developed world. The average age of cars on the road in Australia is ten years. By comparison, the average age of the British vehicle fleet is 6.7 years and in the European Community and USA it's eight years. Older cars are generally more polluting than newer models.

So if you can't drive it, what do you do with your beat-up old car? Dumping is illegal, so recycle the car instead. In Australia, over 500,000 vehicles reach the end of their life each year, and we currently recycle nearly three-quarters (by weight) of the components of our old cars. However, the industry hopes to lift this to 100%.

Take your old car to your nearest auto parts recycler, preferably one that is an Auto Parts Recyclers Association of Australia (APRAA) accredited recycler. Many auto parts recyclers are happy to collect cars that are unregistered or not in running order from homes in their local area. Parts and components in good condition can then be reconditioned if need be and reused in the maintenance and repair of similar models. Scrap metal will also be easily recycled.

When they dismantle your car, they will also 'de-pollute' the vehicle, ensuring that potentially polluting materials, such as waste motor oil, are appropriately dealt with. In particular, car airconditioners often use the refrigerant HFC-134a, which has a very high global warming potential of 1300 (that is, a greenhouse impact over 100 years that is 1300 times that of the same amount of carbon dioxide by weight). It's important that these materials are handled correctly.

In some cases, you may get cash for your old car. You can also go to an auto parts recycler to buy second-hand parts. Visit the APRAA website at www.apraa.com to search for your nearest dealer.

CURRENT FUELS

It's great to use less fuel in your car, but what about the fuel itself? Petrol, diesel fuel, liquefied petroleum gas (LPG) and compressed natural gas (CNG) are all powering vehicles around Australia. They are all hydrocarbon-based fossil fuels, which are not renewable. Their combustion produces carbon dioxide, carbon monoxide and other greenhouse gases, contributing to global warming. These gases and fine particulate emissions (such as tiny bits of soot) contribute to air pollution, which can lead to health problems, particularly in cities where there is concentrated vehicle use.

Fuel options

The fuels of today are very different from those of 30 years ago. Here's a run-down on what's available.

Compressed natural gas (CNG) allows for a reduction in greenhouse emissions of 30–40% for cars and 10–20% for heavy vehicles compared with petrol, and is a relatively clean-burning fuel. Many public transport buses now run on natural gas, which is the same gas that we use in our homes for heating and cooking. CNG engines can be fitted to cars, just as you can convert your car to LPG. A natural gas conversion for a car costs around $2000–2300. The problem with CNG is refuelling. Australia currently has a limited number of refuelling stations for CNG. Private on-site refuelling is also an option where natural gas is available.

Liquefied petroleum gas (LPG) is available at most service stations. If LPG-fitted cars are kept properly tuned, they emit up to 15% less carbon dioxide and less pollutants than petrol-powered cars, though more than CNG.

LPG conversions cost around $2000, but be sure to get a factory-approved kit and have the conversion done by a reputable tradesperson. Your state automobile association (for example, NRMA, RACV or RAA) can give you advice on LPG conversion. If the reduced cost of fuel is your main reason for converting to LPG, the RACV estimates that the fuel savings will pay for the cost of the conversion after around 50,000 kilometres.

Petrol is the most commonly used car fuel in Australia. Petrol is now available in lead replacement, regular and premium. Premium unleaded is the environmentally preferred variety, but they all contribute to smog and climate change.

Diesel fuel can be used in cars with diesel engines. In the past, diesel cars weren't common in Australia's urban areas, but were more popular in country areas, as diesel engines generally require less mechanical maintenance. With growing interest in climate change and rising petrol prices, diesel's share of the car market has grown, taking advantage of greater efficiency and less carbon dioxide emissions compared with the equivalent petrol cars. However, put simply, diesel is dirty. Diesel engines put out much higher particulate emissions and nitrogen oxides (NO_x) emissions than petrol engines. This is of particular concern in urban areas, where it contributes to smog and is linked with lung cancer and a host of other respiratory illnesses.

Biofuels are plant oil and alcohol fuels that can be blended with petrol to reduce petrol consumption and the pollution associated with it. Biofuels are less polluting and, because they come from plant crops, are renewable. Usually the alcohol ethanol is distilled from fermented plant matter. The best plants for making alcohol are those that contain sugars, starch or cellulose, such as sugarcane, sugar beet, cereals and kelp. Ethanol from sugarcane is already being developed in Australia.

Plant oils, including canola, olive and sunflower oils, can be used to supplement diesel fuel. Biodiesel fuels are being developed in countries such as the USA and South Africa and are a possible transport fuel of the future. Some trials are even experimenting with making biodiesel fuel from waste cooking oil from takeaway food outlets. The unwanted waste oil that once cooked your fish and chips may end up powering your car.

Crops can be grown specifically for the purpose of producing biofuels, but in the long term, the planet's growing population may need this land to grow food crops. Already, US demand for biofuel has pushed up maize prices, to the detriment of some of the world's poorest people. Instead, some biofuel projects get their biomass from the unwanted part of food plants, such as bagasse, the waste fibre from sugarcane.

BUYING A NEW CAR

The size, weight and engine type (and consequent fuel type needed) will all affect the overall environmental performance of your car. The two overriding themes of greener car choices are greenhouse emissions and human health. These 'rules of thumb' generally apply:

- Larger cars are less efficient because they are heavier and therefore need more energy to move them around. Four-wheel drives (also known as sports utility vehicles, SUVs and recreational vehicles),

Greenhouse Grand Prix

Which is the leanest, greenest car? If you challenge a late model 4WD, a mid-sized sedan and a petrol/electric hybrid to a year's worth or 20,000 kilometres of driving, the hybrid comes in at first place, producing the least amount of carbon dioxide emissions, with the SUV in last place.

Over 20,000 kilometres:

Hybrid	produces 2 tonnes CO_2
Petrol sedan	produces 4 tonnes CO_2
4WD	produces 6 tonnes CO_2

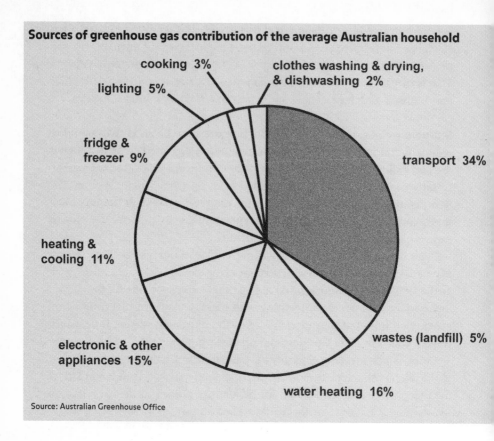

Sources of greenhouse gas contribution of the average Australian household

cooking 3%

clothes washing & drying, & dishwashing 2%

lighting 5%

fridge & freezer 9%

transport 34%

heating & cooling 11%

electronic & other appliances 15%

wastes (landfill) 5%

water heating 16%

Source: Australian Greenhouse Office

being generally larger, are also heavier and therefore more expensive to run.

- Newer cars generally cause less pollution than similar-sized older cars, provided both have been well maintained.
- Cars with lighter paint and interior colours will tend to reflect light, while darker colours absorb light and heat. If you live in a warm climate, choose lighter colours to reduce the need for airconditioning.
- Aim for a five-star car, based on the ratings at the Green Vehicle Guide website. This federal government initiative tests and rates all new cars sold in the Australian market and gives them two scores. The Greenhouse Rating provides a score out of a maximum of ten points, with a higher score indicating lower greenhouse emissions and (typically) lower fuel consumption. The Air Pollution Rating similarly gives a score out of ten, with a higher score indicating the car's exhaust has lower levels of unhealthy pollutants such as carbon monoxide, nitrogen oxides, particulate matter, volatile organic compounds and benzene. The two scores are combined to

give a star rating—the more stars, the better. Visit the Green Vehicle Guide website to compare cars online at www.greenvehicleguide.gov.au.

- You can also compare each model's fuel consumption and go for a more efficient model. The Australian Greenhouse Office has developed a fuel-consumption labelling system for all new cars. The label for any given model shows an estimate of how many litres of fuel that particular model would use to drive 100 kilometres in city conditions. Fuel consumption figures are included in the information provided at the Green Vehicle Guide.
- Consider choosing an engine that runs on a less polluting fuel such as diesel, LPG or CNG.
- If car-pooling is an option, make sure you buy a passenger-friendly car.
- Consider buying one of the latest green machines, such as the electric hybrids now on the market.

Electric, electric hybrid and hydrogen cars

Every motor show trumpets some new mega-eco-car that is super lightweight, runs on peanuts, emits a fragrant rose-scented exhaust and/or can fold into a convenient suitcase. Concept cars are interesting but not a lot of use until they are commercially available.

Now that legislators are convinced that climate change is a reality and that cars are contributing to it, we're finally seeing some of these concepts being developed into viable products, some of which are now available on the market.

Five-star cars

The Green Vehicle Guide is an Australian government initiative to help consumers identify more efficient and environmentally friendly new cars. There are now more than 20 cars that have achieved a five-star rating. The guide's top five performers are:
- Mitsubishi i MiEV
- Tesla Roadster
- Toyota Prius
- smart fortwo
- Honda Insight.
For more information, go to www. greenvehicleguide.gov.au.

Electric hybrid cars use both an electric motor and a petrol engine, working together to power the car. In conventional cars, kinetic energy (the energy of movement) is converted to heat while braking, effectively wasting this energy. The beauty of electric hybrids is that they are able to convert a car's kinetic energy into electricity and store it in a battery for later use. The result is excellent fuel economy and fewer harmful and greenhouse-polluting emissions. They're easy to drive and better for urban air quality.

Sounds like a fabulous concept car that will never be available, doesn't it? In fact, Toyota and Honda have both released hybrid cars to the Australian and New Zealand markets. You could buy one of these green-machines tomorrow. Current models include the Toyota Prius, the Toyota Camry (which is made in Australia), the Lexus GS450h and RX400h, the Honda Civic Hybrid and the Honda Insight.

When only the electric motor is running, hybrids (and all electric cars, for that matter) are virtually silent, so drivers need to be aware that they can't rely on engine noise to announce their presence to pedestrians and other cars. A little toot on the horn can be a useful warning when, for example, coming out of a concealed driveway.

There are also a number of plug-in electric hybrids becoming available overseas, and these are expected to enter the Australian market in the near future. In addition to the features of electric hybrids, these cars can be plugged into household electricity sockets so that they can fully charge with grid electricity.

It's worth mentioning that there's several hybrid myths circulating, some possibly started by die-hard guzzler lovers who relish the opportunity to stick the boot into smug hybrid drivers. The Hummer is not greener than a hybrid (despite what some poorly fact-checked media stories may say); hybrids generally don't have to be plugged into a power socket; the batteries stand up well to the test of time (despite the gloomy predictions of naysayers), and Toyota has a program in place ready to recycle the batteries. But also bear in mind that owning a hybrid doesn't mean that you can drive as much as you want, and it doesn't earn you eco-credit points that allow you to take long showers, buy the biggest TV screen on the market and turn on the airconditioning full blast. Buy it to reduce your emissions and fuel consumption, not just to salve your conscience.

Fully electric cars use only an electric motor to spin their wheels. They get their power from batteries that are recharged by being plugged into the electricity grid. A small number are on the road in Australia, including the Mitsubishi i MiEV, the Tesla Roadster sports car and other cars that have had conversions. They are typically recharged in private garages. Overseas, various recharging options and infrastructure are being trialled, such as roadside plug-in recharging stations incorporated into parking spots, or battery-swap stations, where drained batteries are exchanged for fully charged ones.

They are even quieter than hybrids, posing a risk to pedestrians. Some car companies are trialling vehicle sound systems.

Electric vehicles are a breath of fresh air for urban air quality. Because there is no internal-combustion engine, they do not emit polluting gases to the local environment. However, that doesn't necessarily mean that they are completely clean and green. Their environmental impact is linked to the source of electricity used to recharge the car. If it's electricity from brown coal, then the car is still highly greenhouse polluting. So electric vehicles are a greener choice when they're combined with GreenPower or photovoltaic solar panels.

The hydrogen car Hydrogen is arguably the ultimate fuel of the future. It burns with oxygen to produce energy, and the by-product is not some scary carcinogen but simple water.

The difficulty is in storing the hydrogen. Hydrogen occurs naturally as a gas. As a fuel, it needs to be liquefied and stored in a large, heavy, pressurised and well-insulated tank, which doesn't leave much room in the boot for luggage or shopping. This is the hurdle to overcome before hydrogen-powered cars become a reality on the road.

ALTERNATIVES TO USING CARS

Obviously, the best option is to try to reduce our reliance on cars in our daily lives. Urban planners, employers and individuals can all do a range of things that will help the environment by limiting the use of cars.

What urban planners can do

Civil engineers and urban planners can do a lot for the environment by being creative and 'thinking outside the square'. Here are some ideas that the environmental organisation Greenfleet suggests:

- Improve public transport systems.
- Provide bicycle lanes on popular bicycle routes.
- Provide express lanes on freeways or main arterial roads for cars with two or more passengers or for selected environmentally preferred cars.
- Help fund public transport improvement by charging a levy on inner-city parking.

What employers can do

- Offer public transport vouchers instead of parking in salary packages.
- Make any shower facilities on site available to staff who walk, skate, jog or ride bikes to work.
- Look into the possibilities of online offices and video-conferencing so that some employees can work from home full- or part-time. Many are already looking into this option for parents wishing to work from home after having babies.

What you can do

- Use public transport. A person who lives a five-minute drive from a train station but 40 minutes from work may consider driving to the station

> **Greenfleet: trees for cars**
>
> Greenfleet is an organisation that reduces the environmental impact of transport by planting native trees. These trees offset the carbon dioxide emissions of the cars on Australia's roads.
>
> Greenfleet is funded by donations from people like you who want to neutralise the carbon dioxide emissions of their car. For a tax-deductible sum of $56.95, Greenfleet will plant and care for at least 17 native trees on your behalf. The more you donate, the more trees they will plant. For more information, go to www.greenfleet.com.au.

$ and taking the train the rest of the way into work. This would also save on inner-city parking costs.

- Try two wheels instead of four. Motorbikes, scooters (including the very sexy Vespa), electric scooters and electric bikes are all greener options than a petrol-powered car.

- Consider pedal power. If you haven't been a bike rider since your youth, don't worry: it all comes back to you. Electric bikes, which also require some pedalling, can ease you back into the habit if your fitness levels aren't up to the slopes in your area. Organisations like Bicycle Victoria (www.bv.com.au) and Bicycle NSW (www.bicyclensw.org.au) have programs to support riders who are under-confident or need to brush up on their skills. There are also products that make bicycles more versatile, like saddle bags that sit over the back wheel, or toddler trailers. You can even get folding bikes.

- Move closer to work or to a house with better access to public transport. This is one of the benefits of renting: you can more easily move house when you change jobs.

- Offset some of the emissions of your car by planting trees to help combat the greenhouse effect.

- Share a ride wherever possible. Organise a car pool with workmates. You can make a car pool fun by turning one trip a week into a regular breakfast date. If you're a parent, organise a car pool to school with the parents of your child's classmates.

- Working from home is worth investigating. Although you may miss the social aspects of the office environment, there are some fringe benefits to the home office. At home you can have a bad hair day, work in your pyjamas or play old Duran Duran CDs and no one will know.

→ **ASK TANYA**

I'm considering buying either a new electric-powered bike or a conversion kit for my existing bike, so that I would ride more often and for longer trips. But I'm concerned with just how environmentally friendly and economical this is, considering the costs per kilometre, battery materials and disposal etc. There also seems to be some debate about what batteries are most suitable. Is it worth it?

Anon.

Electric bikes win hands down when compared economically and environmentally with petrol cars, but they suit some lifestyles and family needs better than others. One life-cycle assessment of transport options from MIT reported that 'electric bicycles use less than 10% of the energy required to power a sedan for each mile travelled', taking into account the energy use of fuel production, infrastructure, maintenance and manufacturing, as well as that of driving or riding. Running cost estimates of electric bikes are around 1–3 cents per kilometre, while the RACQ puts the running cost of the current top-selling car (the Holden Commodore 3.0L V6 automatic sedan) at 11.61 cents per kilometre.

There's been a bit of negative hype surrounding electric vehicle batteries, which I suspect has been perpetuated by those who don't want to give up gas-guzzlers. In Australia, car battery recycling is a success story (around 97% are recycled), and the Australian Battery Recycling Initiative is running pilots with the aim of safely recycling all types of domestic dry-cell and vehicle battery. The website www.batteryrecycling.org.au will keep you up to date on the progress of this program and give details of drop-off points for bike (and other) batteries.

Electric bike enthusiast Mike Rubbo, who writes the cycling blog www.situp-cycle.com, recommends choosing lithium batteries that have a guarantee of at least a year. He also recommends e-bikes with detachable battery packs that can be brought inside the house to recharge overnight, particularly if you don't have an enclosed garage or similar spot for recharging. Mike also advises against ordering a bike over the internet. Buy locally so that you can test drive the e-bike, and get it from a reputable dealer who can also provide local servicing. You get what you pay for, so be prepared to pay $2000 plus. Solar photovoltaic recharging options are also available.

Mike thinks e-bikes are wonderful: he lives in a hilly area and probably wouldn't bike-ride without the help of an e-bike.

TRAVEL

Weekends and holidays offer a chance to escape the daily grind and enjoy the delights of recreation and travel. The natural environment offers us fun, inspiration and pleasure, whether we're taking a short day trip or a long holiday. When we're out and about, it's important that our relationship with nature is one of mutual respect, not one of domination and exploitation. We may love our planet, but let's not love it to death by destroying it in the name of fun.

Tips for greener travel and tourism

Don't remove pieces of the natural environment. In nature, nothing goes to waste. Ecosystems have uses for all of their components, living, dead and non-living (mineral) alike. Logs from fallen trees provide homes for small mammals, reptiles and insects. Similarly, shells at the beach can provide homes for sea life, as some sea creatures roam from shell to shell. Broken-down shell grit also helps to replenish the sand on beaches and the sea floor.

Never bring home animals or plants collected in the wild. You could be harming wild populations and the natural ecosystem from which they were taken. In the case of animals, you may be separating breeding pairs, or offspring from their parents, possibly causing them great distress. In addition, these animals aren't domesticated and are unlikely to be as easy to care for as common household pet species.

If you find an injured animal, contact local wildlife authorities. Groups such as WIRES (Wildlife Information, Rescue and Education Services) have trained staff and volunteers who can rehabilitate animals that are not too seriously injured with great success.

Stick to the paths in sensitive areas. Some environments, such as coastal sand dunes, are very fragile. Even the daintiest of human feet can damage them. In addition, homes for wildlife can also be fragile and are best not disturbed. For this reason, stick to the tracks and paths in nature reserves and national parks. Also use boardwalks where they're provided through sensitive sand dune areas.

Respect boundaries. Leave any gates as you found them: shut them if they were shut; leave them open if that's how you found them. They may allow or prevent the movement of both wild animals and domestic livestock. If areas of natural vegetation are fenced off and locked, don't try to access them anyway. Many are fenced for good reasons. For example, some catchment areas, where plant life helps to convert rainfall to water supplies, are fenced off to protect the quality of our drinking water. Other areas may have restricted access because they are habitat for critically endangered and fragile species.

Only use motorised water vessels where they're allowed. Motorboats, jet skis and other water-sport equipment are prohibited in some areas as they can disturb some delicate water environments. Make sure you know your local boating laws, as fines can be hefty.

> ## Swimming with sharks
>
> There are a handful of highly dubious tourism operations and activities that offer people the thrill of coming close to dangerous animals, sharks in particular. Operators attract sharks by dropping fresh meat into the water near the tour boat. Tourists watch the feeding frenzy from the boat or sometimes 'swim with the sharks' in protective cages.
>
> Such activities should be avoided. They teach sharks to associate humans and boats with food, ultimately putting other swimmers, divers and sailors at risk. We are partly to blame for some of the shark attacks of recent years.

Use designated launching sites to launch boats. Backing towing vehicles up to the water's edge can damage riverbanks and cause erosion.

If you're going off-road, drive only where permitted. Driving a four-wheel drive does not give you licence to go off-road anywhere you like. Off-roading can literally tear the ground or sand dunes, causing serious environmental damage. There are areas where off-roading is permitted and is relatively safe. These areas are set aside because they've been deemed tough enough to withstand four-wheel drive traffic and are not significant conservation areas. Make sure you have permission if driving on private land or roads. Avoid marshy areas, and only cross streams where a road crosses them or at designated fording sites.

Wash mud and debris off your vehicle and tyres. Wash your vehicle before and after entering different wilderness areas, particularly if you plan to go off the road. This will help to prevent the spread of noxious or invasive weeds.

Keep the carbon footprint of travel in mind. The Intergovernmental Panel on Climate Change estimates that aviation causes 3.5% of human-made global warming, and that this could rise to 15% by 2050, so keep in mind that there are other ways to see the world and factor this into your travel decisions. The greenhouse impact of each form of transport is not just related to fuel consumption. There is also the energy used to make and maintain the various transport vehicles and the infrastructure that supports them. Taking these things into account, the rough averages of the green-house emissions in carbon dioxide equivalents of the main modes of transport over 1000 kilometres are as follows:

Bus or train	150 kg CO_2
Plane	490 kg CO_2
Car	340 kg CO_2

Offset your travel. Whatever way you get there, estimate the emissions of your travel and offset them.

Choose a greener hotel. If you're staying in accommodation rather than camping or caravanning, choose lodgings that have an environmental-management system and policy. Ask what they do to minimise waste, water and energy use. You would be surprised by how many are becoming greener, particularly since the energy and water used to meet their heavy laundering needs contribute greatly to their running costs.

Ecotourism Australia is a non-profit tourism industry organisation that runs the Eco Certification Program, an environmental standards initiative for the ecotourism industry. The program looks at the environmental and social sustainability of tours, attractions and accommodation. They also run the EcoGuide Australia Certification Program, which accredits eco-tour guides. Visit their website at www.ecotourism.org.au to find out more and to search online for certified tours, attractions and accommodation.

Three ways to Byron Bay

In 2005, three Melbourne-based families took a joint holiday to Byron Bay. One family drove the 1650-km distance over two days in an electric hybrid Toyota Prius. The second family accompanied the first in a large four-wheel drive. Both cars needed about two full tanks of petrol to get to Byron Bay, though the four-wheel drive's 90-litre tank was double the size of the hybrid's 45-litre tank. The third family scored some cheap flights online and flew the distance. Here's an overview of the transport costs for the round trip—both environmental and financial.

Family	Number of people	Transport type	Petrol consumption (litres)	Approx. CO2 emissions (kg)	Financial cost ($)
1	4	Hybrid car	180	450	225
2	3	4WD	360	900	450
3	3	Plane	n/a	4850	800

Choose souvenirs carefully. Some souvenirs and local products may be made from endangered plant or animal species, and this is a particular risk for souvenirs purchased from markets or in countries where regulations are poorly policed.

Don't buy the following products unless you're absolutely certain that they are made from an abundant species:

- wallets, purses, boots, belts and bags made from reptile skin
- anything made from spotted or striped cat fur (including jaguar, snow leopard, tiger and ocelot)
- shahtoosh shawls, made from the fur of the endangered Tibetan antelope, which is killed to collect the wool
- anything made from ivory
- products made from tortoiseshell
- coral jewellery or ornaments.

The Australian government Department of Sustainability, Environment, Water, Population and Communities has a useful website at www.environment. gov.au/biodiversity/trade-use/index.html. Information on this site includes approved sources of wildlife for trade, and guides to buying wildlife products, including a separate publication for complementary medicine products.

Consider a volunteering holiday. 'Voluntourism' combines travel and sight-seeing with volunteer charity work and philanthropy. With some tour companies, meals and accommodation are included in the price, but the often life-changing experiences are priceless. Some packages include luxury accommodation; others are more along the lines of 'roughing it'. Your holiday could include helping a community environmental education program in South Africa or empowering underprivileged communities in India.

Hands Up Holidays is one company offering voluntourism packages. There are travel package options suitable for family groups, solo travellers, corporate groups and even honeymooners. Visit www.handsupholidays.com.

Conservation Volunteers Australia (CVA) and the New Zealand Trust for Conservation Volunteers offer a wide range of green volunteering holidays throughout Australia and New Zealand. You will have to pay your own travel expenses and a certain amount to feed and keep you, but it is often less than you would expect to pay staying in private accommodation. For more information, visit www.conservationvolunteers.org.au or www.conservationvolunteers.org.nz.

CVA also offers volunteer opportunities overseas through World Conservation Programs. Projects run for between two and five weeks, and with most projects participants will be part of a team of volunteers. Destinations include the USA, Europe, Central America, South America and Africa. For more information, visit www.conservationvolunteers.com.au/volunteer/world-conservation.htm.

MAKE A COMMITMENT

This week I will...

☐ _____

☐ _____

☐ _____

This month I will...

☐ _____

☐ _____

☐ _____

When I get the opportunity I will...

☐ _____

☐ _____

☐ _____

Notes

_____

_____

_____

_____

_____

_____

14

How to have a green baby

If you're reading this section you're probably either pregnant or you've already had a baby. You may also be particularly concerned about the future of the world for the sake of the child you're bringing into it.

It's not easy being a mum, but with some planning and awareness of your choices it's not any harder to be a green mother. This section looks at some of the alternatives available to new mums (or 'Mr Mums' for that matter) and offers some simple tips for reducing your impact on the environment while raising your child.

Leading up to and during your pregnancy, your biggest environmental concern is the baby's own local environment—your body! This is a time to clean up your body to ensure ideal surroundings for your child during gestation.

In the months prior to conception and during pregnancy, it is important to avoid exposure to toxic metals (lead, mercury, cadmium and aluminium), toxic chemicals and radiation (X-rays, for example). The information outlined in chapter 2 becomes all the more important.

During pregnancy, the baby's cells are rapidly growing and dividing and are highly sensitive to certain chemicals. Your intake of clean air and water, and food that is free from potentially harmful substances is important for the normal development of your child. At particular stages during pregnancy, exposure to certain chemicals and bacteria can cause serious problems for the development of the foetus. The exact times of these stages are not always predictable, so it is best to err on the conservative side and avoid exposure to harmful chemicals for the duration of your pregnancy.

TIPS FOR A LOW-TOXIN PREGNANCY
- Avoid any renovation in your home that disturbs old lead-containing paint (generally used on houses painted before 1970).
- Have any dental work done well in advance of pregnancy and look into the possibility of replacing any amalgam fillings you may have with non-amalgam alternatives.

- If you're decorating the new nursery, talk someone else into doing the hard smelly work. Now is definitely not the time to expose yourself to fumes and volatile organic compounds (VOCs). Milk your 'delicate' situation for all it's worth! Remember that painted and renovated surfaces can continue to emit indoor air-polluting gases even after the paint and glue have dried, so use low-fume or environmentally certified paints and renovating materials. Avoid using solvents and fume-emitting adhesives.
- Avoid hair-colouring products during pregnancy. Most have health warnings on them advising against use by pregnant women.
- Eat organic food where possible and thoroughly wash fresh fruit and vegetables, even those that are organic.
- Avoid fish that typically have high levels of mercury (shark, ray, swordfish, barramundi, gemfish, orange roughy, ling and southern bluefin tuna). Limit other fish, such as canned tuna, to one portion per week. For more information about mercury and fish, see chapter 7, pages xx–xx.
- Eat free-range, organic chicken, eggs and meat products. Non-organic livestock grown for meat are often fed synthetic hormones and antibiotics to improve their growth and avoid diseases. Exposure to synthetic hormones can confuse the development of the reproductive organs, while the overuse of antibiotics can and has contributed to the development of antibiotic-resistant 'superbugs'.
- Use the green cleaners outlined in chapter 4.
- Quit smoking (if you are a smoker) and avoid smoky environments. Cigarette smoke is a cocktail of toxic chemicals and combustion particles that are particularly dangerous to both children and pregnant women. Smoking is linked with low-birth-weight babies and increases the risk of sudden infant death syndrome (SIDS).

GREEN FOOD PREPARATION

For newborns the best food environmentally, and the healthiest, is breast milk. It has all the nutrients the newborn needs and is easy to make—you can literally do it in your sleep. It must also be said, though, that not all women can breastfeed and they shouldn't feel guilty or inadequate if they can't. Problems with milk supply, inverted or cracked nipples, having to return to work and many other factors can make a mother unable to breastfeed.

Breast milk is particularly important in the development of a baby's immune system. Children who were breastfed as babies for at

Breast shields

If you're breastfeeding, use reusable cotton breast shields instead of disposable ones to prevent the tell-tale signs of leaking breasts. Most disposable shields are made of cotton and cellulose (a wood fibre) so they can sometimes be irritating, particularly on cracked nipples.

least their first six months tend to have lower incidences of allergies and asthma. They also tend to be healthier, as breast milk carries immunity from the mother to the infant.

A common misconception is that women's nipples are not sterile and are therefore dirty, making breastfeeding unhealthy for the child. Breasts do not have to be sterilised. A simple wash with soap in your normal shower is all they need.

Sterilisation of bottles and teats, however, is an issue. Even breastfeeding mothers will generally start introducing solid foods into a baby's diet at around 4–6 months and may want to sterilise plates, bowls and utensils.

There are two basic methods of sterilisation. Applying heat—either by boiling, steaming or using dry heat—uses a physical method to kill bacteria. Chemical sterilisation kills them using a chlorine solution (usually sodium hypochlorite or 'bleach') that is toxic to the bacteria. The Milton Method is a commonly used form of the chemical sterilisation method. This method can't be used with metal objects, as they tend to corrode in the chlorine solution.

It is always better to avoid using chlorine-derived chemicals. Chemical sterilisation leaves chlorine and dioxin residues, even when rinsed with water. Then disposing of the chlorine-containing water places an extra chemical burden on water-treatment systems. Any concern about the energy used in heat sterilisation is outweighed by the greater concern with dioxins and the environmental health problems they are suspected of contributing to.

To sterilise using heat, the items should be subjected to 95°C heat for around two minutes. Immersing the items in boiling water, or using a steam-sterilising machine or a microwave sterilisation kit, can do this. However, take care with boiling and microwave sterilising: both can melt some plastic objects.

As your child gets older, feed him or her healthy food and drink. Use filtered tap water or boiled rainwater (subject to local environmental health requirements) for drinking. Try to prepare fresh, organic foods. There are a growing number of brands of prepared organic baby food, such as Only Organic or Earth's Best Organics, which are available from most supermarkets.

Fresh organic produce is also becoming more readily available. Many of the large supermarket chains now stock a range of organic produce, tea, flour, pasta and other wheat products. There are also a number of online grocery shopping services, but they will only deliver to limited areas. Both Coles Online (www.colesonline.com.au) and Woolworths (www.homeshop. com.au) deliver to most metropolitan areas in Australia's capital cities. Each offers a range of organic fruit and vegetables as well as other grocery items.

The same recycling rules apply to those tiny jars, bottles and food tins as for larger packaging. Rinse them in old washing water and put them in your recycling bin. As with all food preparation, fruit and vegetable scraps can also be composted.

Baby-food jars are great for reusing around the home. They're particularly good for storing small bits and pieces like spare buttons, paper clips and hairpins. If you're stewing food for a baby, why not use a double-boiler and use the steam to sterilise the containers in the top half while the food is cooking.

Dioxins

You may be wondering what dioxins are and what all the fuss is about. 'Dioxins' is the common name for a family of chemicals called polychlorinated dibenzodioxins. There are many types, and several are highly toxic. They do not occur naturally. Rather, they are by-products of the chlorine bleaching process. Small amounts or 'residues' can be found in chlorine-bleached paper products including disposable nappies, tissues, sanitary pads, filter paper and toilet paper. Dioxins are also released when PVC plastic and other chlorine-containing organic materials are burned. They do not break down in the environment.

High doses can cause severe facial acne, lowered immune resistance and liver damage. They are suspected of causing birth defects and increased cancer risk. Dioxins readily dissolve in fatty animal tissue. Effluent discharged from paper mills into waterways brings dioxins to fish and other marine life, harming their health. Humans and other animals eat the fish and the dioxins accumulate in the food chain. While high doses of dioxins are avoidable, scientists are particularly concerned with the suspected effects of low-level, long-term exposure to dioxins, both through the food chain and from residues in bleached products. This is why it's best to reduce our use of chlorine-bleached products.

DOING THE LAUNDRY

It's amazing how one tiny person can produce the same amount of dirty laundry as four adults. This then escalates when solid food is introduced—the adult usually ends up wearing more food than the infant actually eats.

Wash laundry with pure soap flakes or a laundry powder that has a low environmental impact and is phosphate-free. Low-environmental-impact products with no phosphates, optical brighteners or petrochemicals tend to be less irritating to the sensitive skin of babies. For more information about phosphates and optical brighteners, see chapter 4, pages 51 and 54.

If you're worried about the sanitation of clothing, washing in water heated to 65–95°C will kill most of the bugs. Also, sunlight has a sanitising and bleaching effect, so if the weather is warm, sun-dry your washing. If you really feel that you must use a nappy soak, again use one without phosphates, and use non-chlorine bleach.

CHOOSING SAFE TOYS

The first place a baby puts a toy is in his or her mouth. Australia has strict laws about the kinds of materials and paints that can be used to make toys for infants. However, the laws (and how tightly they're adhered to) will not necessarily be the same overseas.

Wooden toys and artefacts brought back from overseas holidays, particularly those bought from roadside markets in developing countries, should generally not be given to babies to play with. You may be giving the baby an item painted with toxic paint. Remember that many of these souvenirs are made to be looked at. They are not designed to be safely munched by a teething infant.

Also be careful with plastic toys, particularly those made from PVC (polyvinyl chloride or vinyl). PVC is one of the worst plastics environmentally, largely because it is produced using chlorine and is very common. Basic PVC is hard, brittle and difficult to use, so stabilisers and plasticisers are added to it to produce finished vinyl products. It is these additives that are cause for concern.

Alarmingly, two stabilisers that were once commonly used are lead and cadmium. Lead poisoning is a well-known public health concern. Cadmium can be more toxic than lead, can cause kidney damage and is linked to cancer. They are still used in some vinyl products, but many countries have laws banning their use in toys. Plasticisers are used in soft vinyl toys like the old favourite rubber duckie. Studies have shown that when children put these soft vinyl toys in their mouths they can ingest plasticisers called phthalates. Phthalates have been linked to kidney damage as well as to problems with the liver and reproductive organs.

Greenpeace International has launched the 'Toxic Toys' campaign to raise awareness of these health risks and to place pressure on governments internationally. The aims of Greenpeace and its allied groups are to have governments around the world ban children's products containing lead and other potentially dangerous additives.

Good old-fashioned wooden toys are still available. The Swedish furniture manufacturer IKEA prides itself on the fact that its toys do not need batteries, they're painted with non-toxic paints and the wood is sourced from sustainably managed timber plantations.

IN THE BATH

See if there are ways that you can reuse your baby's bath water (for example, car, floor and window cleaning).

Pure soap and water will often do just as good a job on the change table as baby lotions. Use simple vegetable-based soaps and shampoos and other plant-based formulations instead of petrochemicals or soaps made from animal fats. Soaps derived from animal fats tend to need more preservative, which can provoke allergies. In fragrances, look for natural essential oils, which will also irritate less than synthetic fragrances.

Be careful when choosing and using baby shampoos. Some 'no-tears' formulas work by adding an antihistamine or similar anti-irritant, which simply masks the irritation rather than preventing it. If you're not sure, it's best to just try to keep the soap out of junior's eyes.

THE GREAT NAPPY DEBATE

Cloth versus disposable: what is the answer? It's the question on every mother's (and sometimes father's) lips, especially when you consider that the average baby has gone through around 6000 (sometimes up to 8000) nappy changes by the time he or she is toilet-trained.

Despite the many more-pressing environmental challenges facing our society, nappy sustainability is the topic I'm most frequently asked about, perhaps because I'm female and a mother. I'm continually shocked by how passionate this debate gets and how judgemental people can be with new parents over their nappy choices. (Yet it's considered socially acceptable to get your electricity from polluting brown coal, despite the wide availability of GreenPower.) My aim in this section is to provide a rational approach to making nappy choices, but I'd also like to make a heartfelt plea to green mums not to berate their friends over their nappy choices. For many people, parenthood may be the first time they think seriously about the future of the planet and consider greener living, for the sake of their child. If we then load them up with guilt over their nappy choices, we risk turning them off before they've had a chance to engage more broadly with environmental issues.

Too often, the nappy debate is oversimplified to a choice between 'wasteful' disposables and reusable cloth. In reality, the environmental considerations of nappy use go far beyond material use and solid waste generation. For many expectant parents, the first nappy challenge is finding credible, relevant information. Dozens of studies have been done around the world to try to weigh up the environmental impact of both cloth nappies and disposables. Many have looked at all stages of their life cycles: the material, energy and water resources used to make them, their production methods, the effect of their use and finally their disposal or reuse. Many were conducted in North America and the United Kingdom, based on assumptions that don't necessarily apply to the Australian situation.

A comprehensive independent life-cycle analysis was commissioned by the British Government Environment Agency, with the findings released in May 2005. This study compared home-laundered cloth nappies, disposables and cloth nappy-wash services and found that 'overall no system clearly had a better or worse environmental performance, although the life-cycle stages that are the main source for these impacts are different for each system'.

Should we take that as gospel for Australia? Well, the United Kingdom has higher ownership of water-saving front-loading washing machines than Australia, and Australia's electricity is, on average, more greenhouse polluting, so the water use and greenhouse statistics from British studies can't be assumed to fit Australian nappy use. Similarly, colder-climate studies assume clothes dryer use. In Australia, clothes dryer ownership varies from 35.1% of Northern Territory households to 61.5% of ACT households, making it difficult to form any assumption about 'typical nappy use'. In short, you will find a lot of information and even scientific studies about nappies on the internet, but they won't always apply to your local situation.

Fortunately, researchers at the University of Queensland have conducted life-cycle analysis of three systems of nappy use in Brisbane conditions: home-washed reusable nappies, commercially washed nappies and disposables. The study focused on four environmental indicators: water use, energy use, solid waste generation and land area required to produce raw materials. Importantly, they highlighted the fact that the major difference between nappy systems is that the user has much more control over the environmental impact of home-laundered nappies and therefore the ability to make choices that minimise the impact. They can choose detergents, how they dry them, the wash cycle temperature and so on. The study found that 'home-washed reusable nappies, washed in cold water in a front-loading washing machine and line-dried … use less energy and land resources, comparable water resources, and produce similar or lower quantities of solid waste, compared with other nappy systems'. Note that it did not give a blanket finding that cloth is better than disposable.

Then there are the new kids on the block. Alternative fibre cloth and disposable nappies are now available, made from fibres such as hemp, bamboo and organic cotton. There are also disposable nappies that are compostable or that have greater biodegradable content. Because of their recent entry into the marketplace, there is very little life-cycle assessment research for these products.

This is certainly one environmental choice that you shouldn't sweat over. However, there are better ways to use each type of nappy. The following is an overview of each method, its pros and cons, and some eco-tips to ensure that your use of nappies takes the smallest possible toll on the planet, whatever type of nappy you choose. And one thing we can all agree on is that the planet will benefit from toilet-training as soon as your child is ready!

Nappy numbers

- In 2009, Australians used more than 2.1 billion nappies.
- In Australia, 95% of babies in nappies wear disposables.

DISPOSABLE NAPPIES

Advantages

Reduced home water and detergent use Disposable nappies do not use the detergent and water that is required to wash cloth nappies. Laundry detergent use is particularly a problem when it contains phosphates, which are bad for the health of our country's waterways. If water supply is a big issue in your area, then disposable nappies may be your only real choice.

Convenience Washing poo- and wee-stained fabric is not high on the list of the nation's favourite hobbies. Disposables are light, not bulky and easy to take with you away from the home.

Reduced wetness Keeping excessive moisture away from a baby's skin can give the baby a better night's sleep. The baby is less likely to be woken up by the discomfort of a wet bottom. This may be a particular issue if you have a baby with difficult sleeping patterns.

Reduced nappy rash On the health front, most disposables are designed to keep moisture away from the baby's skin, which can help to prevent nappy rash.

Disadvantages

Waste disposal problem Disposable nappies are a significant contributor to landfill. Sending waste to landfill is not a long-term solution to any waste-disposal problem. It just leaves a problem for future generations to solve. Disposables end up in landfill, and it's not just their bulk that's the problem. Their contents, particularly when it rains, can leak into the earth along with other landfill leachate and contaminate groundwater. This is becoming less of a problem as landfill sites are increasingly lined and better managed.

Dioxins Chlorine bleach is used in the production of some disposables to give them their whiteness. Dioxins are a by-product of this process and are hazardous for people and the environment. Dioxins are of particular concern, as they are toxic and accumulate in the food chain.

Cost Buying disposable nappies adds a significant cost to your weekly grocery bill. In fact, in 2010 Australians spent over $420 million on disposable nappies at the supermarket.

Resource use Disposables require land use and water to produce the trees for the pulp filling. It takes an estimated 407–829 square metres of land to produce the raw materials needed for each year that a child spends in disposable nappies, compared with 13–40 square metres for home-washed cloth nappies.

Petroleum-derived product Disposables are made partly from plastics, which are an oil-derived product. The planet's oil resources are limited and not replaceable. The extraction of crude oil is polluting and damages ecosystems.

Possible health risk There is some evidence to suggest a link between disposable nappies and male infertility. Wearing disposable nappies tends to raise the temperature of boys' testes. Studies have shown that temperature is important for the normal development of testes and a good sperm count. However, there are other suspected factors in the rising incidence of male infertility, such as sitting for long periods, tight underwear, exposure to synthetic hormones in the environment and inadequate nutrition. Research is continuing in this area.

Tips for using disposable nappies

- Flush the poo before throwing away the nappy. This keeps the excrement out of landfill and reduces the greenhouse gases produced as it decomposes. Urban sewage systems break down excrement in a way that limits the production of methane gas, which has 21 times the greenhouse potential of carbon dioxide. It also reduces the environmental health risk of spreading pathogens found in faecal matter.
- If you choose disposable nappies, consider one of the growing range of eco disposables. Brands include Bambo Nature, Moltex, Ecobots and Nature Babycare. They typically have greater renewable content, and some of their plastics and outer packaging are derived from cornstarch. Favour those that are certified by an independent third-party eco-labelling program and that are chlorine-free. Bear in mind that nothing degrades easily in landfill conditions; the 'benefits' of using these renewable alternative fibres stem from reducing the use of non-renewable petroleum-derived plastics or making use of fibre that's an otherwise unwanted waste product from other manufacturing. For more information on biodegradability, see page 103. If conventional disposables are your only option, look for those that are hydrogen peroxide bleached (as opposed to chlorine bleached).
- As a compromise, you can limit your use of disposable nappies to occasions when convenience really counts, such as going out or going on a holiday.

- The fact that disposables keep wetness away from the baby's bottom is not an excuse to be lazy! Leaving a wet nappy on a baby provides a nice warm and moist environment for bacteria to happily grow. Change disposable nappies regularly to prevent nappy rash and the proliferation of bacteria.
- If you've chosen disposable nappies, then you've taken the high waste and material-use road (as opposed to the high water, detergent and energy-use road). This is *not* a dirty secret and you don't have to feel guilty about it. Instead, do something positive to counterbalance your nappy use. Make a concerted effort to reduce your waste and material use in other areas. Now might be a great time to start a compost bin, if you haven't already.

CLOTH NAPPIES

Advantages
Cost Buying a one-off supply of cloth nappies is a lot cheaper than buying disposables each week. Even using a nappy-wash service is usually a similar price to buying disposables.

Reusable Aside from the obvious day-to-day reuse, cloth nappies can have a life after toilet-training. They can be kept for future babies or used as rags around the home.

Less resource use Because you're using each nappy again and again, you're getting a far better return on the investment of resources needed to make them.

Less waste The contribution of cloth nappies to Australia's landfills is negligible compared with disposable nappies. Using cloth nappies for one baby prevents around 6000 used disposables from being sent to landfill.

Biodegradable Once the nappy is no longer useful, it is biodegradable.

Made from cotton Cotton is good for sensitive skins that may otherwise become irritated by the elastic and plastic in disposables. As a product of agriculture, cotton is also renewable, compared with plastic made from petroleum.

Disadvantages
Water, detergent and energy use Cloth nappies require water, energy to heat the water, detergent, and your time and effort to clean them. However, the extent of this disadvantage can vary, depending on how green your laundry is. If you have a front-loading washing machine that has high

energy- and water-efficiency ratings, you're using green laundry detergents and your electricity comes from GreenPower sources, then cloth nappies becomes a better alternative for your particular household.

Made from cotton The use of cotton is both an advantage and a disadvantage environmentally. Cotton is a crop that has substantial environmental impacts. Non-organic cotton farming generally uses a very high level of pesticides and fertilisers. Such farming methods are harmful to the land and surrounding waterways and can kill plant and animal life. However, nappies made from organic cotton, hemp, bamboo and blends of these fibres are available from environmental and baby product retailers, on the internet or by mail order.

Poor fit Folded cloth nappies don't fit as snugly as disposables and are more likely to leak. They can also be difficult to put on a wiggly baby. Again, there are alternatives on the market to fix this problem, such as fitted nappies.

Dioxins The cotton may be bleached to make it white, before being made into nappies.

Tips for using cloth nappies

- A wide range of more convenient cloth nappies are available from retail baby goods stores or on the internet. These include specially made (and very cute) shaped nappies, usually with elasticised legs and sometimes with Velcro fastenings. These cloth nappies are designed to fit like disposable nappies and they remove the hassle of having to fold flat nappies. However, they can take longer to dry after washing.
- If you've chosen cloth nappies, then you've taken the high water, detergent and energy-use road. You can reduce this impact by using a water-efficient washing machine, greener electricity and greener laundry products, and by drying them on the clothes line or an indoor airing rack.
- You can also counterbalance the environmental costs of cloth nappies by reducing

Washing nappies

Many people like using a two-bucket system for washing nappies. Half-fill one bucket with water and add a little white vinegar and bicarb soda. This is the 'wet' bucket, which the 'number two' nappies go into once the solid matter has been emptied into the toilet. The second bucket is the 'dry' bucket. Put nappies that are only wet into this one. You may wish to put a little bicarb soda in the bottom to absorb smells. Keep lids on both buckets. Once they're full, drain the nappies and wash them with pure soap or a low-impact laundry power. Save the special nappy soak products for nappies with more persistent stains and, even then, only use those that are chlorine-free. They typically use sodium percarbonate (which dissolves in water and yields hydrogen peroxide) as the active ingredient.

your energy and water use elsewhere in the house and by swapping over to green cleaners. Make sure you use only chlorine-free laundry products, and try green cleaning alternatives elsewhere in the house. Now might be the time to get a rainwater tank and swap over to green electricity.

NAPPY ALTERNATIVES AND RELATED PRODUCTS

Nappy wash service

Nappy services supply cloth nappies, collect used nappies, wash them and deliver clean nappies. Individual nappy-wash-service nappies have a shorter life span as they're no longer used once they start to thin or get discoloured, something home washers generally tolerate in their own nappies. Consequently, using a nappy service typically requires greater land use to produce raw materials than home-laundered cloth nappies (25–114 square metres per year of nappy use, compared with 13–40 square metres per year).

When you're shopping around for a service, ask about any environmental auditing or accreditation they may have had done, and what water- and energy-efficiency measures they've undertaken. For example, they may use some solar-heated hot water. Make sure you use a service that uses detergents that are phosphate-free. Also, ask whether they use chlorine in the rinse water. Chlorine bleaching can leave dioxin residues in the nappies.

For those concerned about the hip pocket, nappy services have a similar cost to that of disposables, and are sometimes cheaper, depending on the brand of disposables used for comparison.

Compostable nappies

Tasmanian company Eenee Designs makes a couple of compostable and biodegradable nappy options. The first is 'Weenee Eco Nappies', a range of chlorine-free and plastic-free 'compostable' plant-fibre disposable nappies. Unlike traditional disposables, they work as a two-part pad and pant system. The compostable pads are designed for single use, while the pant that holds them in place is reusable.

More recently, the company developed 'Eenee Compostables', fully biodegradable, shaped disposable nappies that are held in place by a reusable gripper belt, which serves the function of the Velcro tabs on conventional disposables.

Note that 'compostable' doesn't necessarily mean putting them in your home compost

---

Mark a birth with trees

Give the living 'gift of trees' to mark the birth of a baby. When you buy a 'Gift of Trees' from Trees for Life you're effectively sponsoring this South Australian–based conservation group to plant local native seedlings on your behalf. Trees for Life works with volunteers to plant native seedlings to help restore environmentally degraded areas. The 'Gift of Trees' is commemorated with a special certificate. For more information, visit www.treesforlife.org.au.

bin. In fact, the company advises against putting soiled nappies into home compost bins for hygiene reasons. 'Compostable' means they are acceptable for commercial composting.

Flushable plant-fibre nappy liners are also available. These can be used in conjunction with cloth or disposable nappies, making the job of removing excrement and flushing it down the toilet much easier.

Eenee products are available by mail order or through specialty nursery stores. Visit www.eenee.com.au.

Nappy-related products

Plastic nappy bags Avoid buying nappy sacks, the plastic bags for putting soiled nappies in. Instead, keep a couple of supermarket shopping bags from the (hopefully) odd occasion you forget to bring reusable green bags shopping. They can be kept in your nappy bag and used when you are away from home and have a soiled nappy or clothing to take home. They are unnecessary for use inside the home. It doesn't require a lot of effort to take a particularly smelly nappy straight outside to the bin.

Nappy wrappers Automatic nappy-wrapping devices are a wasteful and unnecessary product. They tightly wrap a disposable nappy in an extra layer of plastic, making it harder for the nappy and its contents to break down with reduced greenhouse impact. They also use more non-renewable resources and add more plastic to landfill. Their refills are typically expensive.

Wipes and liners If you are using cloth nappies and already have a washing system for them set up, you may wish to use reusable cloth wipes that can be washed along with the nappies, instead of using disposable wipes. You can make some by cutting up an old towel and overlocking the edges. Alternatively, look for nappy liners and wipes made from renewable and biodegradable materials.

GREEN BABY TO GREEN CHILD

Plant a tree

The best thing you can do for the environment is to teach your child to care for and respect the planet. One way is to plant a tree for your newborn baby or to get an older child to plant one with you. They can watch the tree grow, thinking of it as 'my tree'.

You can even choose trees to attract birds and butterflies. In fact, your own backyard can become an important piece of urban habitat for local native wildlife. Ask your neighbourhood nursery or local council about trees that are native to your local area.

If you don't have a garden but would still like to plant a tree with your child, your local council will probably be able to give you contact details for local tree-planting groups and community garden co-ops.

Grow some food

With our increasingly urbanised population, many children don't know where their food comes from. Involve your children in growing some food at home so that they can learn and appreciate how the environment provides us with food, how plants need water and how food production takes time and effort. Fruit trees, a vegetable patch or even just a couple of potted herb plants on a windowsill can serve the purpose.

Have some good green fun

Give your child positive associations with the planet. We're all more moti-vated by fun and encouragement than by guilt, so give your child (and your-self) the gift of good times spent in the natural environment. Make them aware of the affects humans have had on the environment without wallow-ing in guilt. Involve your child in the natural environment and teach them our connection with it.

Kids love a birthday party, so celebrate Earth Day (22 April) or World Environment Day (5 June) each year with your child and his or her friends by organising a litter clean-up day at a local park followed by a barbecue, or go on a fun nature hike.

Give green gifts

Here are some thoughts to keep in mind when gift-giving:

- Avoid buying unwanted gifts that will just be thrown away.
- If you're buying a battery-operated gift and have the budget, also get rechargeable batteries and a recharger to go with it. This will reduce the amount of battery waste going to landfill.
- Go for quality rather than quantity. Choose engaging and imagination-inspiring presents, rather than a whole lot of useless little knick-knacks.
- Choose gift-wrapping paper and cards made from recycled materials, or wrap children's presents in the colourful cartoon pages from newspapers.
- If you're considering buying a pet for your child, do your research first, as unwanted pets are a huge problem. You might even consider getting a 'recycled' pet from the pound.
- For children aged ten and older, and even adults, another of my books, *Green Stuff for Kids*, is a great green read.

Set an example

As with all aspects of parenting, practise what you preach. As your child grows, he or she learns from you what is normal and acceptable behaviour. When they are old enough, get them to help you with the recycling. If you can demonstrate respect for the environment, your child will adopt your example in the years to come.

Notes

Chapter 1
Local and international electricity tariff information Australian Bureau of Agricultural and Resource Economics (ABARE) report, 'Energy in Australia 2010'.
Black Saturday deaths Queensland University of Technology media release, 'Facing a future of heatwaves: study', 14 December 2010.
Psychology of behaviour change *Fostering Sustainable Behaviour—An Introduction to Community Based Social Marketing* by Doug McKenzie Mohr (New Society Publishers, 2006); 'The psychology of climate change communication', report by Debika Shome and Sabine Marx, Columbia University Centre for Research on Environmental Decisions.

Chapter 2
Air quality-related deaths CSIRO media release, 'Air pollution death toll needs solutions', 2 March 2004.
Factors influencing indoor air quality CSIRO and Bureau of Meteorology report, 'Indoor air project: indoor air in typical Australian dwellings', 2010; Environment Australia report, 'Air toxics and indoor air quality in Australia', 2001.
Endocrine disruptors USA EPA Endocrine Disrupter Screening and Testing Advisory Committee's reports and factsheets; Natural Resources Defence Council's 'Endocrine disruptors' factsheets; Greenpeace International 'Toxic toys' campaign.

Chapter 3
Illumination levels Based on Australian Standard for interior and workplace lighting (AS/NZS 1680).
Paper by numbers Australian Bureau of Agricultural and Resources Economics and Sciences, *Australian Commodity Statistics 2010*.
Chlorine and dioxin information World Bank Group International Finance Corporation report, 'Environmental, health, and safety guidelines: Pulp and paper mills'.
E-waste figures Planet Ark; Clean Up Australia; EPHC, 'National waste report 2010'.
Superannuation asset value Australian Prudential Regulation Authority—Statistics quarterly superannuation performance, December 2010 (issued 10 March 2011)

Chapter 4
Cleaning product market figures Retail Media, 'Retail world annual report 2010', December 2010.
Microbe information *Soap, Water and Common Sense* by Dr Bonnie Henry (UWA Publishing, 2009).

Chapter 5
Value of global clothing market *Well Dressed? The Present and Future Sustainability of Clothing and Textiles in the UK* by Julian M Allwood, Søren E Laursen, Cecilia M de Rodriguez and Nancy M P Bocken; University of Cambridge Institute for Manufacturing, 2006.
Dry cleaning Dave DeRosa, *Out of Fashion: Moving beyond Toxic Cleaners in the Fabric Care Industry*, Greenpeace, USA, 2001.

Chapter 6
Australian cosmetics, perfumes, personal care products and toiletries market value Australian Bureau of Statistics, 8624.0—Retail and Wholesale Industries, Australia: Commodity Sales, 2005–06.

Chapter 7
Grain equivalents Worldwatch Institute.
Food additives and their codes *The Chemical Maze* by Bill Statham; *Additive Code Breaker* by Maurice Hanssen (Lothian, 1986); Calgary Allergy Network; Federation of European Food Additives and Food Enzymes Industries.
Salmon farming information Monterey Bay Aquarium; David Suzuki Foundation sustainable oceans program.
Carbon footprint of food Carbon Trust; Planet Ark; AEA Technology Environment report, 'The validity of food miles as an indicator of sustainable development: final report for Department of the Environment, Food and Rural Affairs (UK)', July 2005.
Water footprint of food Water Footprint Network <www.waterfootprint.org>.

Chapter 8
International waste figures OECD Factbook 2008.
Food waste figures The Australia Institute report, 'What a waste: An analysis of household expenditure on food', November 2009 (note that this figure represents food waste at just the household level); USA EPA, 'Waste not/Want not' program (note that this food waste figure includes waste at agricultural, process, manufacturing, restaurant, retail and household levels).
Population statistics US Census Bureau.
Packaging and recycling 'Planet Ark recycling report' (1997 and 2000 edns); ACI Glass Packaging; Steel Can Recycling Council; Publishers National Environment Bureau; Aluminium Can Group; Australian Liquidpaper Cartonboard Manufacturers Association (ALC); VISY Recycling; Plastics and Chemicals Industry Association (PACIA); Sustainability Victoria; Amcor Recycling; Beverage Industry Environment Council (particularly the BIEC, 'National recycling audit and garbage bin analysis', 1997).

Chapter 9
Emissions from electricity Department of Climate Change and Energy Efficiency report, 'Quarterly update of Australia's national greenhouse gas inventory', September quarter 2010; ABARE report, 'Energy in Australia 2010'.

Controlling heat loss and gain Australian Greenhouse Office publication, 'Global warming, cool it!'; SEDA's 'Live energy smart' brochures; Sustainability Victoria brochures, 'Home cooling hints', 'Home heating hints', 'Sealing out draughts'; GreenHome, Washington, DC, 'Sustainable building, sustainable lives' tipsheets 1 and 2; *Natural Capitalism* by Paul Hawkins, Amory Lovins and L Hunter Lovins (Little, Brown, 1999).

Heating and cooling product guides Sustainability Victoria brochures, 'Choosing a heating system' and 'Choosing a cooling system'; SEDA, 'Guide to energy smart air conditioners'; Department of Climate Change and Energy Efficiency publication, 'Your home: Good residential design guide'; indoor air quality aspects from Asthma Victoria.

Wood fire Environment Australia air quality factsheet, 'Woodsmoke'; Asthma Victoria; Sustainability Victoria brochure, 'Operating hints for wood heaters'.

Stand-by wattage International Energy Association brochure, 'Things that go blip in the night'; Australian Greenhouse Office; ENERGY STAR program; *Choice* report on stand-by wattage.

Home solar energy information Clean Energy Council factsheets; Alternative Technology Association factsheets; Office of the Renewable Energy Regulator.

Home wind energy information Alternative Technology Association publications; Sustainability Victoria guide, 'Victorian consumer guide to small wind turbine generation'.

Chapter 10

The value of Victorian water infrastructure assets Presentation by Tony Kelly, Yarra Valley Water, at the savewater.com.au Water Sensitive Design Seminar, RMIT.

Composting toilet Alternative Technology Association, particularly *The Green Technology House and Garden Book*.

Greywater recycling Based on presentations made at the savewater.com.au Water Sensitive Design Seminar, RMIT; EPA Victoria Information Bulletin, 'Reuse options for household wastewater' from the Domestic Wastewater Management Series (publication 812); 'Greywater' saving water factsheet from Yarra Valley Water; brochure 'Greywater' from Environment ACT; *Sustainable House* by Michael Mobbs (Choice Books, 1998); various Alternative Technology Association reports.

Chapter 11

'Living Soil' figures 'Healthy levels of soil algae lift plant growth' by M. Mallavarapu in *Farming Ahead*, The Kondinin Group, December 2001.

Gardens in bushfire zones NSW Rural Fire Service.

Plants and biodiversity Community Biodiversity Network's 'Earth Alive Action Guide'; Nursery and Garden Industry Australia's 'Flora for fauna' program.

Backyard bullies and invasive species Dr Tony Grice, CSIRO Sustainable Ecosystems, 'The cost of invasive plants', *Plant Talk*, no. 24, May 2001; Kate Blood, Weeds Cooperative Research Centre, 'Hanky panky on the golf course … in broad daylight!', in *Australian Turfgrass Management*, vol. 1.4, August–September 1999; brochure 'Wipe-out weeds' from Victorian Government Department of Sustainability and Environment; National Weeds Strategy.

Rooftop and vertical garden insulation figures Environment Canada Science; *Environment Bulletin*, July 1999; CSIRO and University of Melbourne report, 'Green roofs as an adaptation to climate change' by Dong Chen and Nicholas Williams, February 2009.

No-dig garden advice From waste reduction tips in ACT Government's 'No waste by 2010'; NSW EPA waste and recycling factsheets.

Pest control and pet care Sydney Water, 'Pets, pests and pesticides' project; *Companion Planting in Australia* by Brenda Little (New Holland, 2000); various factsheets on pest control from the Total Environment Centre's 'Toxic chemicals in your home' initiative.

Chapter 12

General sustainable building Numerous interviews with Dr Graham Treloar, Dr Dominique Hes and Dr Robert Crawford, School of Architecture, Building and Planning, University of Melbourne. General sustainable building information from Department of Climate Change and Energy Efficiency publication, 'Your home: Good residential design guide'; Western Power publication, 'Smart home'; HIA GreenSmart initiative; Sustainability Victoria brochures, 'Energy smart house design' and 'Energy smart renovations'; *Sustainable House* by Michael Mobbs (Choice Books, 1998); Alternative Technology Association, *The Green Technology House and Garden Book* (1993).

Victorian experiences of five-star homes Media release from Victorian Minister for the Environment, John Thwaites: 'Overwhelming support for 5 star homes', 17 April 2006.

Material resources RMIT Centre for Design; EcoSpecifier; Good Environmental Choice Australia; Green Building Council of Australia; The Wilderness Society's One Stop Timbershop.

Acknowledgements

Thank you to my husband Andrew, daughter Jasmin and son Archer for their love, support and patience (in other words, for putting up with me while I was writing this book) and to my family and friends for similar reasons—Mum, Dad, Auntie Grace, Daryl, Jill, Graham, Tanya, Marina, David, Chris, Nick, Sandra, Brian, Nicole and Vicki. Thank you also to the extended Ha, Treloar, Durrant, Vick, Bedggood, Welch and RCI families.

Thanks broadly to the teams at Planet Ark, Keep Australia Beautiful, Sustainability Victoria, the Living Smart program (especially Colin Ashton-Graham) and ABC's *Catalyst*—I've learnt so much from working with you and appreciate and admire the work and commitment you give to the environmental cause and science communication. Thank you also to the local councils and the Mums who have been involved with my 'Mama Green' workshops, the readers and viewers who sent their questions to *G magazine* and *Can We Help?* and any one who has shared their questions and concerns with me—you continue to show me what is important to you and how best to work with you, and your enthusiasm and care helps keep me inspired and motivated.

I'd also like to acknowledge the broader environmental science community, particularly my late brother-in-law Graham Treloar. I'm always mindful that the things I write about today were first discovered, investigated or researched by our clever and curious scientists and our society is indebted to them.

It takes a huge amount of work to turn a manuscript into a book—a lot of which I didn't have to do! Thank you to the team at Melbourne University Publishing—Louise Adler, Sally Heath, Foong Ling Kong, Jacqui Gray, Cathy Smith, Georgie Bain, Olivia Blake and the extended team. Thanks also to designer Guy Ivison, editor Lucy Davidson and illustrator Andrew Treloar.

A huge pile of thanks goes to my manager Phill McMartin, particularly for your advice, guidance and frequently used listening skills! Thanks also to the team at Claxton Speakers.

Finally, a heart-felt thank you to the past, present and future readers of *Greeniology 2020*—in particular those who heed the advice and put these words into action. You are our hope for the future.

Index

house: building new, 227–42; buying existing, 227, 242–3; energy systems, 170–7; renovating, 243–6
house envelope, 234–5
household waste, 109–11; hazardous, 117–18
humectants, 78
humidity, 15, 18–19
humus, 204
hyaluronic acid, 89
hydroelectric power, 167, 168
hydrogen energy source, 168

indoor air quality, 11–19, 233
induction cooking, 155
insect control, 55–9
in-slab floor heating, 132–3, 135
insulation, 124, 129, 235–6
investing ethically, 40
irradiated foods, 93–4

junk mail, 108

kerosene heaters, 11, 139
kitchen, 23; cleaners, 49, 53, 54; cooking appliances, 54–7; refrigerators, 124, 149–50; waste water, 189, 191; water use, 182, 183, 188
Kyoto Protocol, 242

La Niña, 180
labels *see* eco-labels and schemes
ladybirds, 215
landfill, 103, 105, 107, 168
landscaping, 5, 30, 195, 218–21
lanolin, 77
laundry, 158, 270; products, 48, 51–2, 55; waste water, 189; water use, 182, 187; whitegoods, 150–3
lawns, 202–3
lead, 245, 271
leaf litter, 211
lecithin, 78
LED downlights, 27, 158, 160, 162
lemon juice cleaner, 46
lighting: artificial, 158–62; bulbs, 163; choosing, 27–8, 161; energy use, 124; fluorescent, 117, 158, 159–60, 163; garden, 161, 218; incandescent, 158–9, 160, 162; office, 25, 26–8;

solar, 157–8, 161
limonene, 53–4
litter, 115–17, 162, 248
living room, 132, 163, 164, 231, 237
LPG (liquefied petroleum gas), 122, 123, 254
lux meters, 27, 38

make-up, 83
manchester, 68–9
manure, 93, 204, 209, 222–3
mattresses, 12, 119
mealy bugs, 215
meat, 94–5, 268
mercury, 16, 268
methane gas, 103–4, 105
microwave ovens, 156
mink oil, 89
mobile phone recycling, 32–3
moisturisers, 86
mosquitoes, 55, 58, 215
mothballs, 64
moths, 58
motor oil recycling, 252–3
mould, 13, 14, 18, 49, 72
mousetraps, 59
mulch, 192, 204, 210–11
musk, 89

nappies, 272–9
National House Energy Rating Scheme, 229
native plants, 194, 195, 196, 197–8, 199, 203, 262
native wildlife, 115, 116, 117, 194, 195, 198–9, 262
natural gas, 122, 123, 178, 254
natural pollutants, 13
newspapers, 110, 112, 113, 204, 210
no-dig garden, 210
non-renewable energy sources, 122

odour removal, 46, 48
oestrogenicity, 77
oestrogens, 77, 78, 89
off-gassing, 14
office, 23–4; consumables, 31–5; educating colleagues, 39–40; energy use, 24–30, 35; equipment, 17, 28–9; furniture, 35; GreenPower, 37–8; heating and cooling, 25–6; lighting,